ADVANCING URBAN SUSTAINABILITY *IN* CHINA *AND THE* UNITED STATES

PROCEEDINGS OF A WORKSHOP

Franklin Carrero-Martínez, Jennifer Saunders, and Emi Kameyama,
Rapporteurs

Committee on Advancing Urban Sustainability in
China and the United States:
A Workshop

Science and Technology for Sustainability Program

Policy and Global Affairs

The National Academies of
SCIENCES · ENGINEERING · MEDICINE

THE NATIONAL ACADEMIES PRESS
Washington, DC
www.nap.edu

THE NATIONAL ACADEMIES PRESS 500 Fifth Street, NW Washington, DC 20001

This activity was supported by the National Academies of Sciences, Engineering, and Medicine's George and Cynthia Mitchell Endowment for Sustainability Science. Any opinions, findings, conclusions, or recommendations expressed in this publication do not necessarily reflect the views of any organization or agency that provided support for the project.

International Standard Book Number-13: 978-0-309-67721-9
International Standard Book Number-10: 0-309-67721-1
Digital Object Identifier: https://doi.org/10.17226/25794

Additional copies of this publication are available from the National Academies Press, 500 Fifth Street, NW, Keck 360, Washington, DC 20001; (800) 624-6242 or (202) 334-3313; http://www.nap.edu.

Copyright 2020 by the National Academy of Sciences. All rights reserved.

Printed in the United States of America

Suggested citation: National Academies of Sciences, Engineering, and Medicine. 2020. *Advancing Urban Sustainability in China and the United States: Proceedings of a Workshop.* Washington, DC: The National Academies Press. https://doi.org/10.17226/25794.

The National Academies of
SCIENCES · ENGINEERING · MEDICINE

The **National Academy of Sciences** was established in 1863 by an Act of Congress, signed by President Lincoln, as a private, nongovernmental institution to advise the nation on issues related to science and technology. Members are elected by their peers for outstanding contributions to research. Dr. Marcia McNutt is president.

The **National Academy of Engineering** was established in 1964 under the charter of the National Academy of Sciences to bring the practices of engineering to advising the nation. Members are elected by their peers for extraordinary contributions to engineering. Dr. John L. Anderson is president.

The **National Academy of Medicine** (formerly the Institute of Medicine) was established in 1970 under the charter of the National Academy of Sciences to advise the nation on medical and health issues. Members are elected by their peers for distinguished contributions to medicine and health. Dr. Victor J. Dzau is president.

The three Academies work together as the **National Academies of Sciences, Engineering, and Medicine** to provide independent, objective analysis and advice to the nation and conduct other activities to solve complex problems and inform public policy decisions. The National Academies also encourage education and research, recognize outstanding contributions to knowledge, and increase public understanding in matters of science, engineering, and medicine.

Learn more about the National Academies of Sciences, Engineering, and Medicine at **www.nationalacademies.org**.

The National Academies of
SCIENCES • ENGINEERING • MEDICINE

Consensus Study Reports published by the National Academies of Sciences, Engineering, and Medicine document the evidence-based consensus on the study's statement of task by an authoring committee of experts. Reports typically include findings, conclusions, and recommendations based on information gathered by the committee and the committee's deliberations. Each report has been subjected to a rigorous and independent peer-review process and it represents the position of the National Academies on the statement of task.

Proceedings published by the National Academies of Sciences, Engineering, and Medicine chronicle the presentations and discussions at a workshop, symposium, or other event convened by the National Academies. The statements and opinions contained in proceedings are those of the participants and are not endorsed by other participants, the planning committee, or the National Academies.

For information about other products and activities of the National Academies, please visit www.nationalacademies.org/about/whatwedo.

PLANNING COMMITTEE ON ADVANCING URBAN SUSTAINABILITY IN CHINA AND THE UNITED STATES: A WORKSHOP

Karen Seto (NAS) (Chair), Frederick C. Hixon Professor of Geography and Urbanization Science, Yale School of Forestry and Environmental Studies, Yale University
Robert Cervero, Professor and Chair, Department of City and Regional Planning, University of California, Berkeley
Frances Colón, Chief Executive Officer, Jasperi Consulting
Susan Hanson (NAS), Distinguished University Professor Emerita, School of Geography, Clark University
Jiahua Pan, Director, Institute for Urban and Environmental Studies, Chinese Academy of Social Sciences
Yong-Guan Zhu, Professor, Biogeochemistry and Environmental Biology; Director General, Institute of Urban Environment, Chinese Academy of Sciences

Project Staff

Franklin Carrero-Martínez, Senior Director, Science and Technology for Sustainability Program; Theme Lead, Global Policy and Development
Emi Kameyama, Associate Program Officer, Science and Technology for Sustainability Program
Melissa Franks, Senior Program Assistant, Science and Technology for Sustainability Program (until February 2020)

Consultant

Jennifer Saunders, Consultant writer

Preface and Acknowledgments

Both the United States and China are experiencing major urban challenges due to rapid population growth and other factors. For instance, in the United States, the urban proportion of the population has grown to over 80 percent in the last decade. Demographers forecast that it is expected to reach about 89 percent by 2050. During this same period, China will experience even more dramatic urbanization: Its urban population will increase from about 50 percent to about 78 percent. This growth is taking place as cities face growing challenges in multiple fronts, including, among other things, climate change, water and energy shortages, pollution, and aging infrastructure. As the growth of Chinese and U.S. cities continue to accelerate, it is important for the scientific community to support research that will further collective understanding of the interconnections between the natural and built environments and how they interact with society.

On December 16, 2019, the National Academies of Sciences, Engineering, and Medicine's Science and Technology for Sustainability program, in collaboration with the Chinese Academy of Sciences, convened a one-day public workshop on urban sustainability in China and the United States. The workshop focused on the intersection of urban climate change mitigation and adaptation, urban health, and sustainable transportation, including green infrastructure and urban flooding in both countries. The workshop was made possible by financial support from the National Academies George and Cynthia Mitchell Endowment for Sustainability Science, whom we thank for its support.

This Proceedings of a Workshop was prepared by the workshop rapporteurs as a factual summary of what was presented and discussed at the workshop. The planning committee's role was limited to planning and convening the workshop. The statements made are those of the rapporteurs and do not necessarily represent positions of the workshop participants as a whole, the planning committee, or the National Academies of Sciences, Engineering, and Medicine. We wish to extend sincere thanks to all the members of the planning committee for their contributions in scoping, developing, and carrying out this project.

This Proceedings of a Workshop was reviewed in draft form by individuals chosen for their diverse perspectives and technical expertise. The purpose of this independent review is to provide candid and critical comments that will assist the National Academies of Sciences, Engineering, and Medicine in making each published proceedings as sound as possible and to ensure that it meets the institutional standards for quality, objectivity, evidence, and responsiveness to the charge. The review comments and draft manuscript remain confidential to protect the integrity of the process.

We wish to thank the following individuals for their review of this proceedings: Chris Hendrickson, Carnegie Mellon University (retired); Catherine Ross, Georgia Institute of Technology; Yan Song, University of North Carolina, Chapel Hill; and Kongjian Yu, Peking University. Although the reviewers listed above provided many constructive comments and suggestions, they were not asked to endorse the content of the proceedings nor did they see the final draft before its release. Responsibility for the final content rests entirely with the rapporteurs and the National Academies.

Franklin Carrero-Martínez, Senior Director
Science and Technology for Sustainability Program

Contents

1 **INTRODUCTION**   1
Workshop Context and Organization, 2
Introductory Remarks, 4
Overview and Goals of the Workshop, 5

2 **CURRENT LANDSCAPE FOR SUSTAINABLE URBANIZATION RESEARCH AND PRACTICE IN THE UNITED STATES AND CHINA**   7
U.S. Perspective, 7
Chinese Perspective, 10
National Science Foundation Perspective, 12
Discussion, 13

3 **URBAN SUSTAINABILITY RESEARCH ACTIVITIES AT THE UNIVERSITY LEVEL**   15
Urban Development in China, 15
Sustainability Challenges at the University Level, 17
Role of Universities: A Synthesis of Recent Research Projects, 19
Discussion, 21

4	ARCHITECTURE, URBAN DESIGN, AND SUSTAINABLE CITIES IN CHINA AND THE UNITED STATES	23
	Nature-Based Solutions and Performance, 23	
	Microclimate Regulation in Cities, 26	
	Happy Cities: Maximizing Human Well-Being Through Urban Design, 29	
	A Climatic Perspective, 30	
	Discussion, 34	
5	DATA AND EARTH OBSERVATION FOR DECISION MAKING	35
	Big Earth Data in Support of the Sustainable Development Goals, 35	
	Data Gaps and Urban Expansion in China, 38	
	Lessons from the Earth from Space Institute's Inaugural Symposium, 39	
	Discussion, 44	
6	ADDRESSING KEY INTERSECTING ISSUES IN URBAN SUSTAINABILITY	45
	Circular Economy and Green Infrastructure, 45	
	People-Centric Design for Sustainable Cities, 46	
	Benefits Of Cross-Sectoral Approaches for Urban Sustainability, 48	
7	THE WAY FORWARD: FUTURE NEEDS AND OPPORTUNITIES	51

APPENDIXES

A	Workshop Agenda	55
B	Biographies of Speakers and Moderators	59
C	Registered Workshop Participants	73

1

Introduction

In both the United States and China, significant urban challenges and changes to the urban landscape are occurring, largely due to rapid population growth in these areas. While in the United States, the urban proportion of the population has been over 80 percent for the last decade, it is projected to reach about 89 percent by 2050 and the numbers in cities is expected to continue to increase by about 100 million between 2010 and 2050.[1] During this same period, China will experience even more dramatic and rapid urbanization: Its urban population will increase from about 50 percent to about 78 percent. By 2050, China will be the first country with 1 billion urban dwellers.[2]

This rapid expansion in cities in both the United States and China has and will continue to place pressure on environmental, economic, and social systems. Urban areas are facing growing challenges from climate change, water and energy shortages, pollution, and aging infrastructure. As the growth of Chinese and U.S. cities continues, it is critical to support research

---

[1] United Nations, Department of Economic and Social Affairs, Population Division. 2018. World Urbanization Prospects: The 2018 Revision, Online Edition. File 3: Urban Population at Mid-Year by Region, Subregion, Country and Area, 1950–2050 (thousands). Available at https://population.un.org/wup/Download. Accessed March 10, 2020.

[2] United Nations, Department of Economic and Social Affairs, Population Division. 2018. World Urbanization Prospects: The 2018 Revision (see table I.5 on page 18). Available at https://population.un.org/wup/Publications/Files/WUP2018-Report.pdf. Accessed March 10, 2020.

that will further understanding of the interconnections between the natural and built environments and how they impact human health in urban areas.

Universities have been a driver in research around urban issues and as such have begun to advance urban sustainability research and data to date. To support innovation at the university level, it is essential that the science community collectively assess the state of urban sustainability research in China and the United States while strengthening and enhancing partnerships to effectively meet current and future sustainability challenges. There is also a need to foster coordination and collaboration for strengthening the science–policy interface by promoting dialogues between scientists and policy makers to adopt best practices.

In November 2018, National Academy of Sciences (NAS) President Marcia McNutt visited China for the first time in her official role. As part of this visit, the National Academies of Sciences, Engineering, and Medicine's Science and Technology for Sustainability (STS) program and the Chinese Academy of Sciences organized a one-day workshop relating to urban sustainability in Beijing. The goal of the visit was to begin to explore some areas where the U.S. National Academies and the Chinese Academy might develop collaboration. At the same time, the trip provided an opportunity to develop links to other parts of the science and technology community in China.

WORKSHOP CONTEXT AND ORGANIZATION

To further elucidate some of these issues and build upon current partnerships, an expert committee under the STS program, in collaboration with the Chinese Academy of Sciences, organized a one-day public workshop on urban sustainability in China and the United States, held on December 16, 2019. The workshop focused on the intersection of urban climate change mitigation and adaptation, urban health, and sustainable transportation, including green infrastructure and urban flooding in both countries.

The workshop discussions were directly related to several of the United Nations' (UN's) Sustainable Development Goals (SDGs), described as "the blueprint to achieve a better and more sustainable future for all." These goals address the "global challenges we face, including those related to poverty, inequality, climate change, environmental degradation, peace and justice."[3]

---

[3] United Nations. 2019. About the Sustainable Development Goals. Available at https://www.un.org/sustainabledevelopment/sustainable-development-goals. Accessed April 16, 2020.

FIGURE 1-1 United Nations Sustainable Development Goals.
SOURCE: United Nations, 2019. Communications materials. Available at https://www.un.org/sustainabledevelopment/news/communications-material. Accessed April 16, 2020.

According to the UN, the 17 SDGs are "all interconnected, and in order to leave no one behind, it is important that we achieve them all by 2030"[4] (see Figure 1-1). Workshop discussions were focused on those most relevant to SDG 3 (good health and well-being); 6 (clean water and sanitation); 9 (industry, innovation, and infrastructure); 11 (sustainable cities and communities); 13 (climate action); and 17 (partnerships for the goals).

After opening remarks (summarized below), the agenda was organized in several sessions; this proceedings follows the organization of the workshop. The framing remarks provided context for the current landscape for urbanization research and practice in the United States and China (Chapter 2), followed by three panel sessions. The first session examined urban sustainability research activities at the university level in both countries (Chapter 3). Researchers from both countries discussed their work and ongoing collaboration on urban sustainability issues, including the importance of academia in fostering innovation around urban challenges. A second session focused on architecture, urban design, and sustainable cities

---

[4] United Nations. 2019. About the Sustainable Development Goals. Available at https://www.un.org/sustainabledevelopment/sustainable-development-goals. Accessed April 16, 2020.

in China and the United States, and included a discussion of nature-based solutions (Chapter 4). A third session addressed data and earth observation for decision making, including discussion of the major advances in these areas for supporting decision making in urban planning (Chapter 5). The session that followed included breakout discussions where participants addressed issues related to the intersection of urban climate change mitigation and adaptation, urban health, and sustainable transportation, including green infrastructure and urban flooding in China and the United States (Chapter 6). A final discussion gave participants the opportunity to reflect on what new efforts might be needed to advance the field, including identifying next steps (Chapter 7). The appendixes include the agenda, biographical sketches of committee members and presenters, and a list of participants (Appendixes A through C).

Presenters and participants had expertise in urban sustainability, green infrastructure and sustainable transportation, urban human health, climate change adaptation and mitigation, and urban flooding, among other areas, and represented China and the United States on these issues. The workshop began with introductory statements by leaders of the U.S. National Academy of Sciences and the Chinese Academy of Sciences.

INTRODUCTORY REMARKS

Marcia McNutt, president of the U.S. National Academy of Sciences, welcomed workshop participants and thanked those who had traveled to the workshop for their participation and support. The rapid pace of urbanization in the United States and China, as well as other countries around the world, presents one of the biggest challenges faced today, said Dr. McNutt. There will be 11.2 billion people living in cities by the end of the century. Now is the pivotal time, she said, to envision what a city can and should look like and to consider how they can be sustainable and producing what they need to be resilient.

To address these issues, research on the interconnectedness of these challenges is needed, she said, to understand the human, not just the natural, dimensions of sustainability. This will require global partnerships. The UN SDGs are oriented around research and innovation; they are needed to address key urban sustainability issues and include measurable targets for achieving progress by 2030. The SDGs can energize the global community around urban sustainability challenges and further partnerships, she said.

Yaping Zhang, vice president of the Chinese Academy of Sciences, added that urban sustainability is not a national but a global challenge, particularly in developing countries. Between 1950 and 2015, there was a shift from 29.6 percent globally living in cities to 53.9 percent, with an average of 140,000 people moving to cities each day. In China, 10.6 percent of the population lived in cities in 1949, and 60.3 percent of the population lived in cities by 2019.[5] The uncontrolled expansion of urban areas in China has been of great concern and is at the forefront of research and development at the national level, Dr. Zhang explained. The Chinese Academy of Sciences has launched a research program to address urban challenges and the results of the current workshop will help to inform that work. Both of these efforts have and will encourage the sharing of policy and knowledge related to urban sustainability challenges moving forward.

OVERVIEW AND GOALS OF THE WORKSHOP

Planning committee chair Karen Seto, Frederick C. Hixon Professor of Geography and Urbanization Science in the Yale School of Forestry and Environmental Studies at Yale University, stated that there are a number of sobering statistics around the changing urban environment, as described above. Demographers forecast that the urban population is expected to continue to increase in the United States by about 100 million between 2010 and 2050. During this same period, China's urban population will increase from about 50 percent to about 78 percent, said Dr. Seto, adding that China is seen as a leader in how it is building its cities. However, the size of the urban population has massive implications for the environment, economy, and social systems, including challenges from climate change, water and energy shortages, pollution, and aging infrastructure.

Dr. Seto explained the goals of the workshop are to support research to advance understanding of the interconnections between the natural and built environments and impact on human health; increase partnerships and memoranda of understanding (MOUs) between U.S. and Chinese universities to accelerate research; and address the urgent need to assess the state of urban sustainability research in China and the United States and enhance partnerships. She added that a key element is to support research that pushes the science forward and, in turn, accelerates research that is being conducted in universities to allow for knowledge sharing.

---

[5] Knoema. 2019. China—Urban Population as a Share of Total Population. Available at https://knoema.com/atlas/China/Urban-population. Accessed March 9, 2020.

The workshop objectives include a number of topics related to urban sustainability in China and the United States focused on the intersection of urban climate change mitigation and adaptation, health, and sustainable transportation, including green infrastructure and urban flooding. Urban areas are at the confluence of these topics and cannot be viewed in silos or individually, she said. She noted the key issues to be discussed during the workshop include:

- Reviewing the current landscape for sustainable urban development policies and practices in China and the United States;
- Highlighting urban sustainability research at universities in China and the United States, including the most significant outcomes of research in an interdisciplinary context;
- Identifying research needs and knowledge gaps toward sustainable urbanization; and
- Discussing effective mechanisms for strengthening the science–policy interface and adopting best practices to address current and future urban sustainability challenges in both countries.

2

Current Landscape for Sustainable Urbanization Research and Practice in the United States and China

With the current pace of urban population growth in both the United States and China and the significant challenges cities are facing as a result, it is critically important to support research in sustainability that will inform decision making and practice. This research must be grounded in the interconnections between the natural and built environments. Speakers from both countries provided the framing remarks focused on the current landscape for sustainable urbanization research and practice that set the stage for the workshop.

U.S. PERSPECTIVE

Deb Niemeier, the Clark Distinguished Chair in Energy and Sustainability in the Department of Civil and Environmental Engineering at University of Maryland, began by noting that taking on the near-term challenge of research related to urbanization as it relates to climate change is critically important. The term "sustainability" has become complicated in meaning; it is often used to define action on things that are clearly not sustainable (e.g., clean coal). She argued that mitigating greenhouse gases (GHGs) should be the primary focus of research on sustainability over the next 5 years. It is nearly universally accepted by scientists that the Earth is moving past a tipping point on climate change with only a few years to act responsibly, Dr. Niemeier said. If we want sustainability, we need to focus on reducing GHGs, she said, and China and the United States produce the largest share of cumulative GHG emissions. There are approximately 8 to 10 years to determine how a projected 98 million more people, most of whom will be residing in cities, will experience their day-to-day choices and living environment.

Dr. Niemeier discussed the breakdown of GHG emissions by sector in the United States (see Figure 2-1), noting that transportation and energy sectors are currently the most significant producers. To address GHG emissions, technological innovation is needed, but while new technologies are deployed, attention must be paid to the structural inequalities that can be hardened or emerge through policy changes. There are important feedbacks between the design and implementation of new technologies, the way policies are introduced to manage energy, and finally who has access to and at what cost these technologies are available. Dr. Niemeier stated that the urban GHG challenge must tackle three research pillars simultaneously: (1) the vast expansion of renewables; (2) the development of a smarter, more flexible energy grid; and (3) the significant increase in products that use electricity.

It is particularly important as renewable options for energy increase to learn how to manage the energy they generate more nimbly. What is needed, she continued, is more system flexibility, new and modern power systems, and new frameworks that can change the way that energy markets operate.

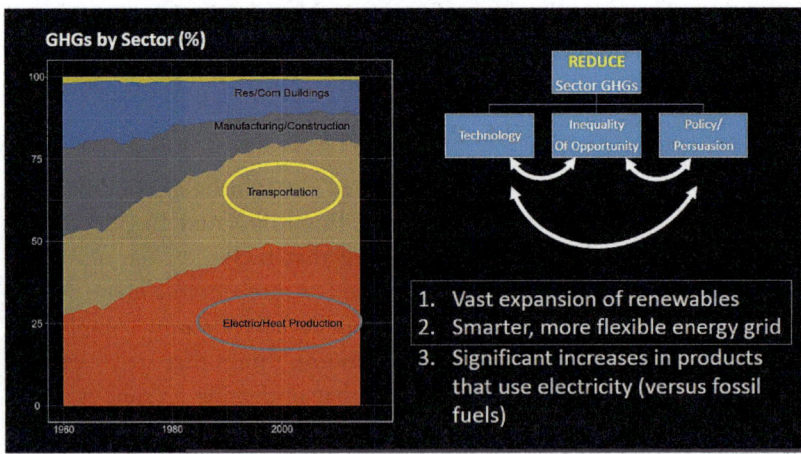

FIGURE 2-1 GHGs emitted by sector in the United States and opportunities to mitigate.
SOURCE: Deb Niemeier, Presentation, National Academies of Sciences, Engineering, and Medicine, December 16, 2019, Washington, DC.

Centralized facilities and those distributed to the power grid are almost always fossil fuel driven, with a constant supply of fuel and supply lines that requires enormous concentrations of capital. Renewable energy, however, is commercially viable and can be found everywhere. Thirty-one U.S. states could meet their entire energy needs with renewables. With renewables, however, there are questions about power generation and ownership.

Dr. Niemeier discussed examples of innovations in cities. For example, many cities are exploring how to implement net zero districts. One pilot in New York City has its energy technologies linked, in which buildings share and exchange heat and electricity. This type of net zero innovation can drive policies that move toward lower costs and lower energy use management. Policy changes are needed that drive the modernization of power plants and utilities as well, said Dr. Niemeier. Utilities understand this need—a recent Black and Veath survey[1] found that 73.7 percent of utilities indicated that distributed energy systems would shape their modernization investments going forward.

Returning to the feedback between new technologies and inequity, Dr. Niemeier added that as new technologies and management strategies emerge, it will be important to attend to issues surrounding inequality and energy policy. Families with the least means pay disproportionally more for their energy use, and most utilities lack data on those who are at the lower income brackets. New research on technologies must concurrently examine mechanisms for addressing current and future structural inequalities as well as ways to integrate more progressive policies around renewals.

Transportation is the least diversified sector and yet produces 28 percent of GHGs, said Dr. Niemeier. As we think about changes to transportation, she urged, we must decarbonize the sector. China has the largest number of electric vehicles on the road today.[2] The emphasis in the United States on improved travel behavior modeling is short-sighted, she said. In

---

[1] Black & Veatch. 2019. Strategic Directions: Smart Utilities Report. Available at https://www.bv.com/sites/default/files/2019-11/SDR_SmartUtilities_2019.pdf. Accessed April 28, 2020.

[2] In 2018, around 45 percent of electric cars on the road globally were in China, while the U.S. accounted for 22 percent of the global fleet. Although the majority of China's electricity comes from coal, driving an electric car is still far more environmentally friendly than driving a gasoline-burning vehicle. See International Energy Agency. 2019. Global EV Outlook 2019: Scaling-up the transition to electric mobility. Available at https://www.iea.org/reports/global-ev-outlook-2019; and Smith, R. 2018. The surprising truth behind the world's electric cars. The World Economic Forum. Available at https://www.weforum.org/agenda/2018/03/electric-cars-are-still-coal-powered. Both accessed March 9, 2020.

the near term, there will be virtually no impact on GHGs without a reduction in vehicle miles traveled (VMT). Visionary changes—which occur very rapidly—are needed to reduce VMT, she said.

Dr. Niemeier noted that the need to address GHG is essential, and the transition to a low carbon environment is problem driven and immediate. Theoretical models will not help move the needle on this issue. According to her, research that turns the dial on the street, gets people out of their cars, and pushes society beyond the current depth of renewable technology adoption will be critical. The urban environment is the best chance for changing the energy future of nations, but the focus must be on pathways to reduce GHGs by harnessing collective intellect and wisdom, she concluded.

CHINESE PERSPECTIVE

Wei-Qiang Chen, professor of resources and urban sustainability at the Institute of Urban Environment, Chinese Academy of Sciences, described urban sustainability challenges in China, including from his personal experience leaving the country and returning to witness the impact of significant growth in cities. China's urbanization since 1978 was driven by and also drove the growth of almost everything. This included gross domestic product (GDP), city sizes, infrastructure, and urban inhabitants. Dr. Chen noted that nearly 600 cities were developed between 1978 and 2018, and the proportion of Chinese living in cities increased from 18 percent to 59 percent during this same period.[3] The scale and size of these cities reflect that of Manhattan versus the smaller and less urbanized character of many other U.S. cities.

In addition to this growth, there are efforts under way to connect cities with modern infrastructures, such as public transportation and high-speed rail, as well as enable the population to achieve a modern lifestyle (air conditioning, refrigeration, e.g.) within three to five decades, said Dr. Chen. However, there are limits to this growth. For example, nearly all economic activities and cities are located in about 40 percent of China's land area, the ecological-geographical limit, also known as the Hu-Huanyong line, which puts significant environmental pressure on this region (Figure 2-2). China is approaching an environmental limit, as

---

[3] Farrell, K., and H. Westlund. 2018. China's rapid urban ascent: An examination into the components of urban growth. *Asian Geographer* 35(1):85–106.

CURRENT LANDSCAPE

FIGURE 2-2 Real-time population distribution approximated by the heatmap of Tencent users in China, as of December 10, 2019.
SOURCE: Wei-Qiang Chen. Presentation, National Academies of Sciences, Engineering, and Medicine, December 16, 2019, Washington, DC, based on https://heat.qq.com. Accessed April 16, 2020.

it produces over 50 percent of the global cement, iron, and aluminum. The aging population in China is another key issue and presents different challenges related to urbanization.

Dr. Chen added that the growth over the past three to four decades has resulted in significant negative lock-in effects that must be addressed in the coming decades. These include the over and unnecessary expansion of urban land use; the problem of shrinking and even empty cities; and the inappropriate design of infrastructure. As an example of an infrastructure design challenge, urban surfaces are over cemented, resulting in flooding and the break of biogeochemical cycles in urban and peri-urban areas. Improper decisions around transportation systems, linked mostly to the pursuit of cars plus high density, and a lack of parking spaces, are contributing to these problems.

Dr. Chen added that there are new challenges to the health of both cities themselves and people living in cities due to this unprecedented growth. These challenges include a dramatic increase in solid and plastic waste, higher-calorie diets, and less physical activity, among others. The country is taking steps to address these challenges, including ministries designed to address key sustainability issues and efforts to promote the circular economy.

NATIONAL SCIENCE FOUNDATION PERSPECTIVE

Linda Blevins, deputy assistant director of the Directorate for Engineering at the National Science Foundation (NSF), described the work of the Foundation on urban sustainability issues, including efforts to collaborate with China. She pointed out that many of today's urban sustainability research efforts focus on individual cities and communities, often addressing transitions within single infrastructure sectors (for example, water or energy) and individual sustainability outcomes (for example, resilience) in a limited number of case study cities. To answer new, complex research questions about urban sustainability may require some people to take a different approach to fundamental research—an approach that is multiscale, deeply integrates disciplines, and involves collaboration among multiple cultures and stakeholders.

Dr. Blevins said that NSF funded a $4.5 million project for research and education in 2012 on low-carbon sustainable cities, focusing on cities in Asia. This project was distinctive in many ways: its twin focus on research and education; deeply interdisciplinary and complex subject matter; and unique, important, and high-quality international research and education opportunities it provided. All of these features made this project a good fit for NSF investment, she said. Another initiative funded by NSF in 2015 included convening 40 people from various organizations to identify challenges to sustainability, systems-based approaches, and fundamental principles and natural processes in the built environment.

In 2016, NSF and the National Natural Science Foundation of China funded a $1 million project on integrated systems modeling of the food-energy-water nexus. Focusing on Beijing and Detroit, the project included creating a modeling framework to identify areas of research and evaluate the consequences of various technology and policy scenarios with the goal of identifying ways to better manage the food-energy-water nexus.

The above is only one example of the collaboration between NSF and the National Natural Science Foundation of China, who have jointly funded 27 projects. These projects resulted from joint proposals over the past 5 years and Dr. Blevins said she hopes that this important collaboration will continue. In 2020, another competition to identify research projects may be jointly funded, she added.

Dr. Blevins pointed participants to a January 2018 report produced by NSF from its external advisory committee, *Sustainable Urban Systems:*

Articulating a Long-Term Convergence Research Agenda.[4] The report provides a vision for a new interdisciplinary science area on this topic, she said.

DISCUSSION

During the discussion session, participants asked speakers if and how cities are tracking progress toward urban sustainability goals. Dr. Chen noted that several Chinese ministries or institutions have been trying to set up goals or indicators for monitoring and measuring progress. However, there is no widely acknowledged systemic goal yet. In addition, there have been pilot projects on examining how to reduce emissions. Another participant noted that in the Intergovernmental Panel on Climate Change (IPCC) report, the authors reviewed 100 climate plans from cities, and while many were ambitious, not one had a single measure for tracking emissions.

Participants also highlighted the generational tension or the interest of younger generations to drive change in urban sustainability. How can this energy be harnessed, several participants asked. Several participants also discussed policies that could be used to break out of behavioral lock-in, including how social media can play a role.

---

[4] National Science Foundation. 2018. Sustainable Urban Systems: Articulating A Long-Term Convergence Research Agenda. Available at https://www.nsf.gov/ere/ereweb/ac-ere/sustainable-urban-systems.pdf. Accessed March 9, 2020.

3

Urban Sustainability Research Activities at the University Level

The first panel session of the workshop focused on urban sustainability research activities at a university level, moderated by planning committee member Susan Hanson, Distinguished University Professor Emerita at Clark University. Universities are often the primary driver of research around urban sustainability issues. Academic research informs understanding of the extent of these challenges; universities also serve as an incubator for innovation as well as a resource for solutions that can drive change.

URBAN DEVELOPMENT IN CHINA

Jianming Cai, full professor at the Institute of Geographical Sciences and Natural Resources Research, Chinese Academy of Sciences, reiterated previous speakers by describing the growth in urban areas in China over the past 40 years. Challenges as a result of this growth and urbanization include urban-rural disparity; heavier regional environmental pressure; traffic congestion; land use inefficiency; and higher living cost, particularly impacting socially vulnerable groups. China has also experienced dramatic socioeconomic change during this period.

Dr. Cai said that China tries to address climate change issues as the country is experiencing intensified flooding, heat waves, water shortages, agricultural production in the hinterlands magnified by the high concentration of population, and assets in urban areas. Dominance along China's east coast is increasing in terms of gross domestic product (GDP) and in population size. Policy attempts to re-orient the focus of demographic growth to the interior have had limited success, given the overwhelming momentum and advantages seen in eastern cities, particularly the three mega urban clusters of the Greater Bay Area, Yangtze River Delta, and Beijing-Tianjin-Hebei.

While the urban population in China will reach nearly 80 percent by 2050, it will begin to decline into the 2050s, Dr. Cai stated (see Table 3-1). This population decline will be highly variable across cities. The urban population is also aging rapidly. He pointed to a need to improve demographic forecasting and knowledge of population dynamics to inform an understanding of labor force participation and productivity of older demographic cohorts.

China's new urban economy is becoming more consumption driven, said Dr. Cai. In response to this, Chinese planners need to ask what new types of urban modules are needed to support China's economic restructuring, he suggested. This would result in better alignment between city building and rapid economic restructuring. Future cities will be "living platforms," characterized by mixed land use, a higher proportion of floor space in residential, and lifestyle services.

Dr. Cai added that China has been successful over the last four decades in building urban road and rail infrastructure, but the emphasis needs to

TABLE 3-1 Forecast of Urban Population Increments in China between 2020 and 2050

Year	Population Increments
2020–2025	81.5 Million
2025–2030	61.3 Million
2030–2035	41.8 Million
2035–2040	23.8 Million
2040–2045	8.6 Million
2045–2050	–0.9 Million

SOURCE: Jianming Cai, Presentation, National Academies of Sciences, Engineering, and Medicine, December 16, 2019, Washington, DC, based on United Nations, Department of Economic and Social Affairs, Population Division. 2018. World Urbanization Prospects: The 2018 Revision. File 19: Annual Urban Population at Mid-Year by Region, Subregion, Country and Area, 1950–2050 (thousands). Available at https://population.un.org/wup/Download. Accessed April 16, 2020.

shift to quality. The country should also adopt a stepwise urban renewal paradigm for creating a more socially inclusive community, said Dr. Cai. For example, there is a need for the organic development of communities with more retrofitting, infilling, and preservation of cultural heritage, vegetation, and other resources as population growth declines and the demand for urban quality increases.

The emphasis should turn to soft infrastructure investment such as portable pensions and health care insurance, improving community livability and urban services delivery, environmental quality, and housing quality, said Dr. Cai. Current funding mechanisms for China's cities are unsustainable. To encourage cross-jurisdictional integrated investment in urban areas, new mechanisms such as national matching grants should be considered, said Dr. Cai.

Dr. Cai discussed several ongoing research needs that could be driven at the university level, including examining spatial planning; research on urban clusters; big data, including opportunities for data collection through smart cities; socially inclusive community building; precision poverty reduction; and heritage-based tourism, among others.

SUSTAINABILITY CHALLENGES AT THE UNIVERSITY LEVEL

Luis Bettencourt, inaugural director of the Mansueto Institute for Urban Innovation and professor of ecology and evolution at the University of Chicago, discussed urban sustainability challenges at the university level in the United States, focusing on the experiences of the University of Chicago. He began by noting that the urban sustainability challenges of today will not be solved through engineering innovation; new science and approaches to these problems are needed. These new approaches are being driven in part by an important generational shift that is pushing change, particularly on climate change issues. The Mansueto Institute for Urban Innovation studies the processes that drive, shape, and sustain cities, with the social, natural, and computational sciences, as well as the humanities to support global, sustainable urban development.[1]

Dr. Bettencourt described scaling effects of urban systems to understand how cities are interrelated, including individuals, neighborhoods, cities, and urban systems. He noted that cities transform at a speed that is unprecedented and unsustainable and are reaching a tipping point. In

---

[1] See https://miurban.uchicago.edu. Accessed March 9, 2020.

October 2019, the C40 Cities announced that 30 of the world's largest and most influential cities have peaked greenhouse gas (GHG) emissions, including Chicago.[2] These cities are in desperate need to take action. Global partnerships are critical to addressing this challenge, and it is fundamental to have a partnership with China to move forward, he said.

Universities also play an important role in testing and designing solutions to key urban sustainability issues, Dr. Bettencourt said. The University of Chicago developed its Sustainability Plan Baseline Report in 2016 to implement its sustainability strategy in nine key areas and manage GHG emissions.[3] Key activities at universities include research, demonstration, education, and partnerships. There is a fundamental need for new research and better data and evidence related to urban sustainability challenges. For example, no good urban sensors exist that can help collect data to inform solutions or research on mixed methods at multiple scales. Also, he said, the issue of equity and its role in all of these issues cannot be underestimated.

Dr. Bettencourt stated that universities provide opportunities for expanding education on urban issues, changing the way to train researchers, and interesting younger children in science. New initiatives, such as Environmental Frontiers, provide University of Chicago students a scientific and practical understanding of urban sustainable development.[4] Universities also offer a pipeline of talent for new fields, constantly identifying and promoting the transformation of new ideas, and a network for scholars to engage and develop partnerships globally.

In terms of partnerships, universities offer opportunities to engage on policy and practices related to urban sustainability and can serve as laboratories, providing data collection opportunities. They can inject new ideas into old conversations, contribute to active learning, and are incubators for innovation and research and data collection and design, concluded Dr. Bettencourt.

---

[2] Additional information about C40 cities and greenhouse gas emissions can be found at https://www.c40.org/press_releases/30-of-the-world-s-largest-most-influential-cities-have-peaked-greenhouse-gas-emissions. Accessed March 9, 2020.

[3] The University of Chicago Sustainability Plan Baseline Report. 2016. Available at https://sustainability.uchicago.edu/sp. Accessed March 9, 2020.

[4] Additional information can be found at https://miurban.uchicago.edu/our-initiatives. Accessed March 9, 2020.

ROLE OF UNIVERSITIES: A SYNTHESIS OF RECENT RESEARCH PROJECTS

Yan Song, director of the Program on Chinese Cities and a professor in the Department of City and Regional Planning at the University of North Carolina at Chapel Hill, discussed how universities can contribute to urban sustainability through a synthesis[5] of her recent projects related to urban form and travel behavior, air quality, urban vibrancy, and economic value outcomes. Collaboration is key; in just the past 11 years, she has collaborated with over 380 scholars from Chinese universities.

One study assessed the impact of compact development on travel behavior and tailpipe emission in Mecklenburg County, North Carolina. Examining travel behavior and other factors, such as projections around the number of households and employment, the researchers were able to model emissions in 2050. This research was designed to help inform decision making and represents a method of linking results with practice, a new approach for interdisciplinary research on urban issues.

In another study examining urban air quality, Dr. Song noted that air quality and other population measures were examined in 152 cities, including cities in China. Examining a number of variables, the researchers found that high density and a strong center are associated with fewer air pollutants, while excessive or even chaotic land use mixture is associated with more air pollutants. Dr. Song noted that new approaches to this research included collecting and examining big data through sub-centers to assess exposure to pollutants.

To examine health effects associated with pollutants, Dr. Song explained the researchers used the Disease Surveillance Points (DSP) system, which forms a nationally representative sample of mortality for 2005. The categories are selected from the International Classification of Disease Revision 9 (ICD-9). Cardiorespiratory causes of death are lung cancer, heart diseases, vascular disease, and respiratory diseases. Researchers found that urban form elements (e.g., density, connectedness, and forest/green space) have significant impacts on $PM_{2.5}$ concentration, thus influencing the incidence of cardiorespiratory mortality at the county level (see Figure 3-1).

To assess urban form, urban vibrancy, and economic values, Dr. Song noted that researchers calculated a composite vibrancy index for 65 down-

---

[5] Relevant material includes Yuan, M., Y. Song, Y. Huang, S. Hong, and L. Huang. 2017. Exploring the Association between Urban Form and Air Quality in China. *Journal of Planning Education and Research* 38(4):413-426. https://doi.org/10.1177/0739456X17711516.

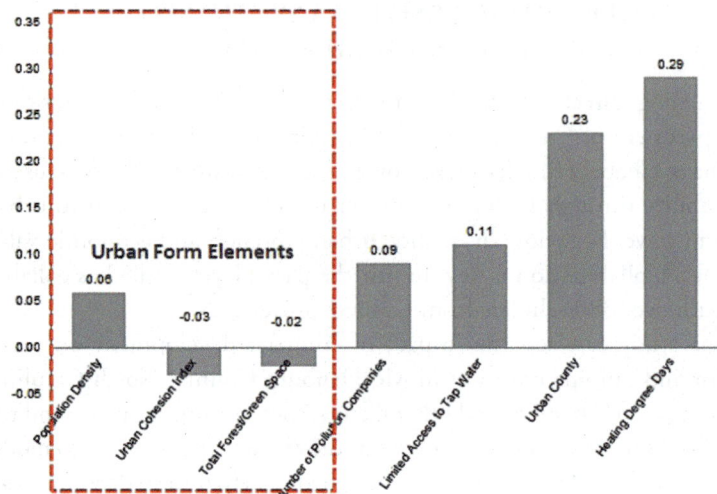

FIGURE 3-1 Standardized effects of urban form and other factors on cardiorespiratory mortality rate.
SOURCE: Yan Song, Presentation, National Academies of Sciences, Engineering, and Medicine, December 16, 2019, Washington, DC.

town areas using standard scores (mean values of zero and scores in standard deviation units) from several measures. With the same unit of measure, researchers were able to create one overall index of vibrancy, retaining standard deviation as the unit of measure. For example, Portland has high live and work, walk and bike scores, public transit use, good street pattern, and smaller block size, while Los Angeles' main contributors to its vibrancy are job density and social diversity with a high percentage of non-English speaking households.

Dr. Song added that collaborative research with China has been vital to understanding and assessing these key urban sustainability issues, developing data, and asking the difficult questions around whether cities are really vibrant and sustainable. Linking policy to research is critical to making decisions about urban issues. Smart governance will require an examination around how to link policy to research in urban areas and examine how effective those policies are, including around communication.

DISCUSSION

Speakers discussed the role of academia in inspiring change around urban sustainability issues. They also discussed the idea that large and small cities face different problems; for example, large cities have sophisticated economies but less space capita, while small cities have less severe environmental problems. As both face a spectrum of problems, what universities are doing to help address these issues must also be fluid and flexible, a participant suggested.

The need for and use of big data were discussed. One participant noted that researchers are collecting panels of data to examine how sustainability interventions are working and to see if they can isolate the effects. There are, however, concerns with big data, and a need to use caution in terms of how these data are used and being validated.

Participants also discussed the role of urban sustainability in addressing inequality, noting that almost every city has a sustainability plan in place that mentions this issue. While data exist that can help assess inequities in the United States, one participant commented, this remains a significant challenge for researchers working on urban sustainability.

4

Architecture, Urban Design, and Sustainable Cities in China and the United States

The second panel of the workshop focused on urban design and landscape architecture as critical components for urban sustainability. The use of nature-based solutions; microclimate regulation for connecting science, policy, and design; and efforts to maximize human well-being through urban design in both countries were discussed. Robert Cervero, professor emeritus of city and regional planning at University of California, Berkeley, moderated the session. He began by noting there are key questions around how to design built and natural environments in urban areas; the science-policy interface is critical; and it is important to consider who should be taking the lead to implement solutions.

NATURE-BASED SOLUTIONS AND PERFORMANCE

According to Kongjian Yu, professor of urban and regional planning and landscape architecture at Peking University, nature-based solutions will be critical to addressing urban sustainability challenges (see Box 4-1). Most cities in China are facing multiple environmental and ecological challenges including floods and urban inundation, drought, groundwater reductions, water pollution, habitat loss, and air pollution, and single-goal minded, industrial technology-based gray infrastructure is not sufficient to solve these interconnected problems. Dr. Yu noted that looking to ancient wisdom as a source of nature-based and holistic solutions can inform decisions about how to address urban challenges and secure ecosystems services.

> **BOX 4-1**
> **Three Key Challenges in Practicing Nature-Based Solutions for Urban Systems**
>
> Dr. Kongjian Yu described three key challenges in practicing nature-based solutions for urban systems, including:
>
> (1) Planning challenges—Space and land are limited and it is critical to consider how to identify and plan the most efficient ecological infrastructure to produce ecosystems services. Identifying ecological security patterns across scales based on analysis of ecological processes can help to address this.
>
> (2) Design and engineering challenges—Nature-based solutions and ancient wisdom might not be efficient nor standardized for modern engineering. We need to conduct research on design and engineering to strengthen performance and standardize them for modern practices.
>
> (3) Policy challenges—Policy changes, including changes to values and education around urban issues, are needed. Such changes require rethinking of the way to build cities based on industrial technologies and calls for nature-based solutions.

Over a period of 20 years, Dr. Yu reported having tested and built over 500 projects in 200 cities in China that have integrated nature-based solutions to address key urban challenges at various scales. For example, Dr. Yu described work at Yongning River Park in Zhejiang in 2003, which demonstrated an ecological approach to flood control and stormwater management while offering space for public enjoyment (Figure 4-1).

Dr. Yu described another example in Yanweizhou Park in Jinhua City, Zhejiang Province, using natural mechanisms to clean contaminated water that introduced an ecological embankment to reduce peak flow. This initiative, if implemented comprehensively throughout the drainage, can result in reducing the flow by more than half at the basin's outlet, making the city more water resilient. As a solution for water management challenges,

FIGURE 4-1 Use of a resilient ecological embankment to replace the former concrete flood control infrastructure, Yongning River, Taizhou, Zhejiang.
SOURCE: Kongjian Yu, Presentation, National Academies of Sciences, Engineering, and Medicine, December 16, 2019, Washington, DC.

Dr. Yu described a demonstration project of a "sponge city"[1] in Sanya, Hainan Province, that addresses urban inundations and floods through a constructed wetland inspired by the ancient wisdom of pond-dyke practice in the middle of the city. The project was designed to retain and filtrate stormwater, which solved multiple urban water issues in a symbiotic way.

In another project for Houtan Park in Shanghai, Dr. Yu demonstrated a strengthened nature-based water cleansing process that can clean contaminated river water to nonpotable clean water through a constructed wetland park, suggesting one hectare of this kind of wetland can produce 800 tons of nonpotable water. This nature-based water cleansing design has been replicated in many cities in China, such as Haikou's Meishe River (Figure 4-2).

Dr. Yu added that today more than ever, a paradigm shift in the planning and design of cities is needed to adapt to the changing climate

---

[1] For additional information related to the sponge city concept, see Yu, K. 2017. Green infrastructure through the revival of ancient wisdom. *Bulletin of the American Academy of Arts & Sciences*.35–39.

FIGURE 4-2 Constructed wetlands to cleanse contaminated water, Haikou Meishe River, Hainan, China.
SOURCE: Kongjian Yu, Presentation, National Academies of Sciences, Engineering, and Medicine, December 16, 2019, Washington, DC.

and to address a multitude of urban ecological issues. Such a shift calls for rethinking the way cities are built, moving toward nature-based solutions.

Moving forward, Dr. Yu posed the following questions for informing future research on nature-based solutions to urban sustainability challenges:

- How can we change our mindset (and policy) to move toward nature-based systems?
- How can we fully assess the performance of nature-based solutions?
- How can we standardize nature-based solutions for climate change adaptation and ecological restoration and scale them to meet larger economic and policy needs?

MICROCLIMATE REGULATION IN CITIES

V. Kelly Turner, assistant professor of urban planning at University of California, Los Angeles, focused her presentation on key challenges and potential pathways for developing actionable science to address the problem of rising temperatures in cities. She discussed the disjuncture between regional scale/generalizable science and the practical everyday decisions that municipalities must make to deal with urban heat issues. There are several key barriers to actionable climate science, including incomplete knowledge around urban design and microclimate (see Figure 4-3). Without addressing some of these barriers, responses to these critical issues will continue to be limited, said Dr. Turner.

Other barriers include limitations around climate data (most available data are coarse) and urban heat profiles. Dr. Turner also pointed to a lack of good understanding of how people use space, as well as a general lack of systemic monitoring of urban sustainability interventions. A disconnect between generalizable, regional urban climate science models and the hyperlocal and context-dependent nature of urban design interventions can lead to policy "panaceas," which she defined as a mismatch between environmental benefit narratives and actual performance in situ.

However, urban environments are replete with naturally occurring experiments that can inform research and action on urban issues. Reframing urban design as urban experiments, akin to the role of environmental policy in the adaptive management process, is a productive first step in reconnecting science and design, stated Dr. Turner.

Dr. Turner added that novel approaches, such as microclimate zones with land cover types at tens of meters, may be used as a tool to guide design at the hyperlocal scale. This level of analysis can allow a focus on a limited suite of controlled conditions that determine, for example, the heat outcomes at those scales. While Stewart and Okie's (2012)[2] local climate zones for urban temperature studies provide a typology of urban land cover types at hundreds to thousands of meters, microclimate zones also allow for a move toward a user-centered approach by effectively summarizing field data at scales that are relevant to municipalities.

Dr. Turner also described the importance of maximizing cobenefits in urban design, highlighting the example of an art installation that used solar paint (see Figure 4-4). It has the cobenefit of reducing heat produced in the built environment as well as beautifying the space. The arts can provide creative solutions to urban challenges, said Dr. Turner.

---

[2] Stewart, I. D., and T. R. Oke. 2012. Local climate zones for urban temperature studies. *Bulletin of the American Meteorological Society*, 93:1879–1900. http://dx.doi.org/10.1175/BAMS-D-11-00019.1.

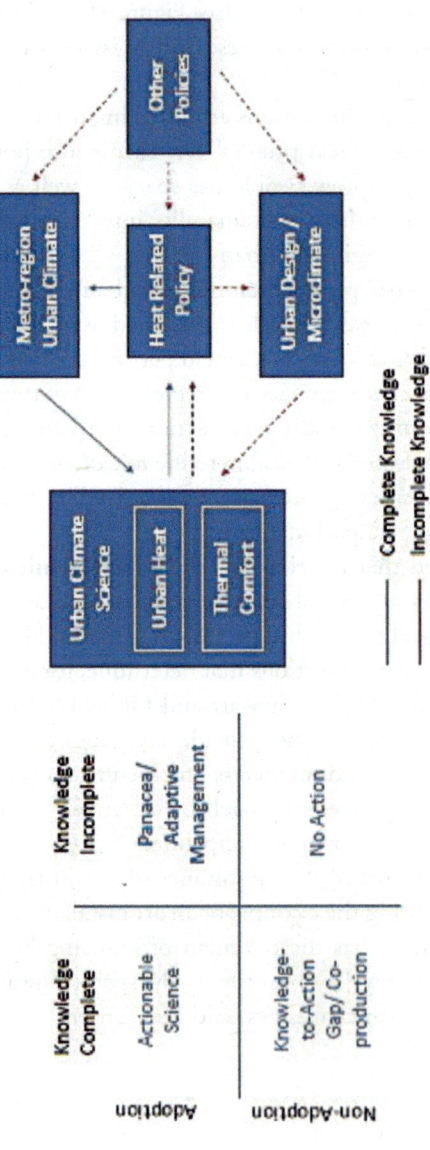

FIGURE 4-3 Barriers to actionable urban climate science.
SOURCE: V. Kelly Turner, Presentation, National Academies of Sciences, Engineering, and Medicine, December 16, 2019, Washington, DC.

FIGURE 4-4 Art installation in South Los Angeles using solar-reflective coating paint.
SOURCE: V. Kelly Turner, Presentation, National Academies of Sciences, Engineering, and Medicine, December 16, 2019, Washington, DC. Thermal imagery created by Ariane Middel, Arizona State University, and photography by Mary Braswell, University of California, Los Angeles.

HAPPY CITIES: MAXIMIZING HUMAN WELL-BEING THROUGH URBAN DESIGN

Yingling Fan, professor in the regional planning and policy area at University of Minnesota, provided additional insight into how urban design can impact human well-being by discussing her research on transportation and happiness in China and the United States,[3] particularly how design can impact or affect emotions. Dr. Fan said that an understanding about how public spaces can evoke emotions can inform future urban design decisions.

Transportation is an emotional landscape, said Dr. Fan. In any U.S. urban area, significant portions of land are devoted to streets, parking, public transit, and sidewalks. Dr. Fan described efforts to study human

---

[3] Fan, Y., R. Brown, K. Das, and J. Wolfson. 2019. Understanding trip happiness using smartphone-based data: The effects of trip- and person-level characteristics. *Transport Findings*. https://doi.org/10.32866/7124. Accessed March 9, 2020.

happiness and travel using an app to measure human behavior. The researchers observed differences between levels of happiness and/or stress based on the modes of travel in Shenzhen and Minneapolis (see Figures 4-5 and 4-6). For example, biking and walking seem to make people happier than using cars in Minneapolis (see green bars), while those greener modes seem to make people less happy than using cars in Shenzhen. There is a need to create a transportation structure that makes people happier when they take greener modes, Dr. Fan stated.

This research examined how various modes of travel and their impact on well-being could be used to inform policy decisions. Dr. Fan also described efforts to examine happiness as it correlates with other factors, for example, having a travel companion, distance and duration, gender, and age, among others. Through this type of work, researchers are able to examine which streets in urban areas are considered risky and which make people happy. Dr. Fan concluded by stating that urban design strategies that focus on cobenefits toward both environmental sustainability and human well-being will have more long-lasting and broader population-level impacts than strategies that merely focus on environmental sustainability.

A CLIMATIC PERSPECTIVE

Matei Georgescu, associate professor in the School of Geographical Sciences and Urban Planning in the College of Liberal Arts and Sciences at Arizona State University, further discussed urban design and architecture, focusing on heat and climate issues. Dr. Georgescu noted that there are twin forcing agents driving climate change in urban environments: greenhouse gasses (GHGs) and the physical built environment. If cities are focused solely on reducing GHG emissions, they are missing half of the problem.

In the urban climate system, there is a balance of incoming and outgoing energy fluxes. Surface energy budgets of urban areas and their more rural surroundings differ because of variability in (1) land cover and surface characteristics, and (2) level of human activity (e.g., how people use energy). Dr. Georgescu described analyses of individual versus total impacts of GHGs[4] and urbanization, including adaptation scenarios, such as the

---

[4] Georgescu, M., P. E. Morefield, B. G. Bierwagen, and C. P. Weaver. 2014. Urban adaptation can roll back warming of emerging megapolitan regions. *Proceedings of the National Academy of Sciences of the United States of America* 111(8):2909–2914.

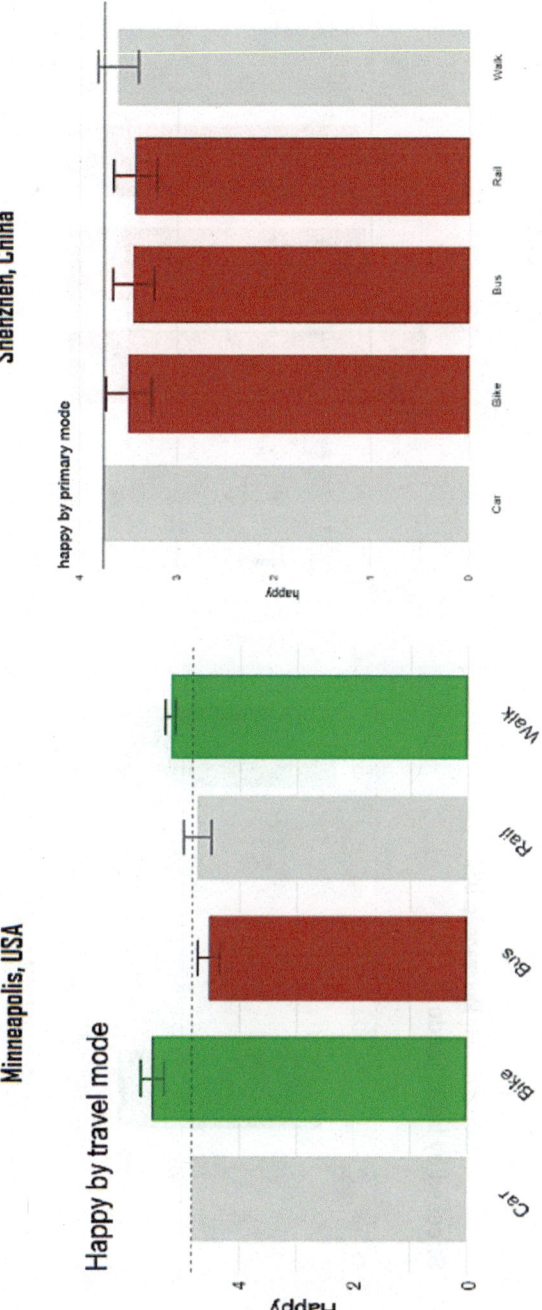

FIGURE 4-5 Measures of happiness by travel mode in Minneapolis, MN, and Shenzhen, China.
NOTE: Adjusted for age, sex, race, employment status, income, family status, general health, life satisfaction, optimism/pessimism, disability, neighborhood characteristics, and trip duration.
SOURCE: Yingling Fan, Presentation, National Academies of Sciences, Engineering, and Medicine, December 16, 2019, Washington, DC.

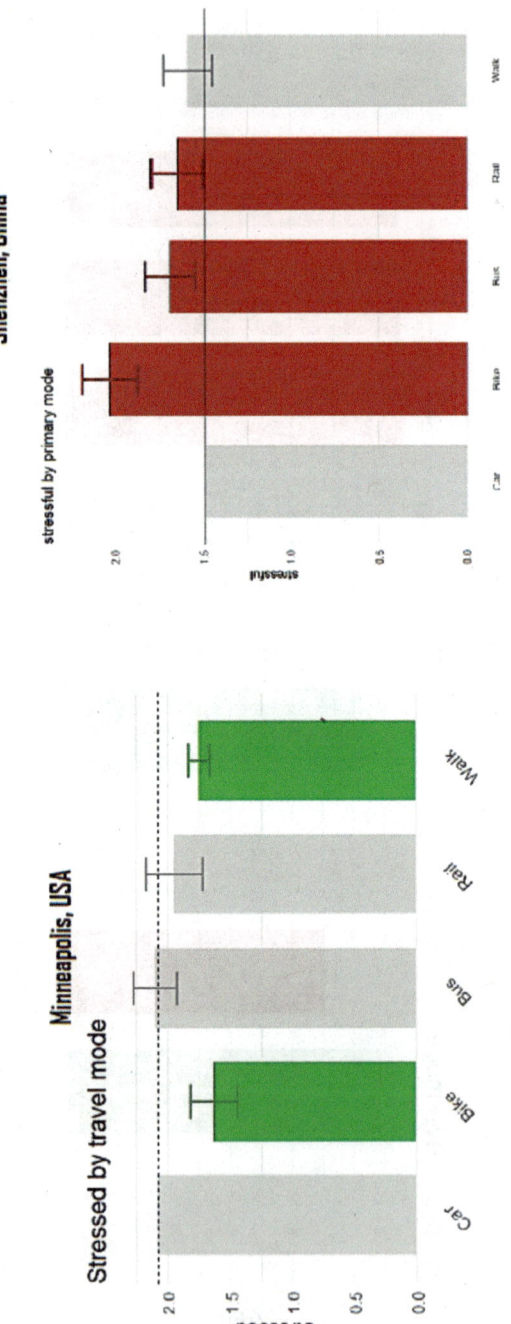

FIGURE 4-6 Measures of the level of stress by travel mode in Minneapolis, MN, and Shenzhen, China.
NOTE: Adjusted for age, sex, race, employment status, income, family status, general health, life satisfaction, optimism/pessimism, disability, neighborhood characteristics, and trip duration.
SOURCE: Yingling Fan, Presentation, National Academies of Sciences, Engineering, and Medicine, December 16, 2019, Washington, DC.

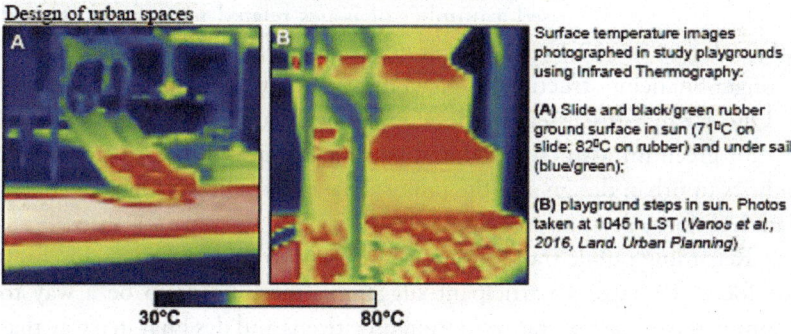

FIGURE 4-7 Example of the effects of heat exposure on playgrounds in Phoenix. SOURCE: Matei Georgescu, Presentation, National Academies of Sciences, Engineering, and Medicine, December 16, 2019, Washington, DC. Reprinted from Vanos, J. K., A. Middel, G. R. McKercher, E. R. Kuras, and B. L. Rudell. 2016. Hot playgrounds and children's health: A multiscale analysis of surface temperatures in Arizona, USA. *Landscape and Urban Planning* 146:29–42, with permission from Elsevier.

dynamic interaction between effects of climate change and urban expansion from the continental U.S. perspective.[5]

Dr. Georgescu noted that heat waves are occurring more frequently and the resulting effects of heat exposure on human health are being seen. Using an example of the impact of heat on urban design in Phoenix, Arizona, Dr. Georgescu noted that the surface temperature images of playgrounds show the potential for health impacts of a heat wave on children in a community (see Figure 4-7).[6] Indoor living environments during these heat waves are also of concern in the United States and, especially, elsewhere.

Dr. Georgescu discussed key knowledge gaps around sustainable urban settlements, including the need for scaling from local to regional levels as well as for the impact of research activities into the *desired outcomes* framework. The scientific community currently does not know what the desired outcomes are or should be, in large part because they are place-based and differ from one geographical location to another.

---

[5] See Krayenhoff, E. S., M. Moustaoui, A. M. Broadbent, V. Gupta, and M. Georgescu. 2018. Diurnal interaction between urban expansion, climate change and adaptation in US cities. *Nature Climate Change* 8(12):1097.

[6] Vanos, J. K., A. Middel, G. R McKercher, E. R. Kuras, and B. L. Rudell. 2016. Hot playgrounds and children's health: A multiscale analysis of surface temperatures in Arizona, USA. *Landscape and Urban Planning* 146:29–42.

DISCUSSION

Participants discussed a number of issues related to architecture and urban design in the United States and China, including the need for a stronger financing structure for green infrastructure and policies to support it. One example discussed was developing mechanisms to fast-track financing for green infrastructure. Several participants also discussed the role of politics in urban design and the importance of engaging citizens to collect data to inform decision making.

Regarding public transportation in China, to date, efficiency has been the focus. Perhaps, a participant suggested, there needs to be a way to examine the impact of transportation on citizens and design it in a way that considers the experience of public transportation.

5

Data and Earth Observation for Decision Making

Jiahua Pan, director of the Institute for Urban and Environmental Studies at the Chinese Academy of Social Sciences, moderated the final panel on major advances in data and earth observation for supporting decision making in urban sustainability planning. The discussion below highlights the presentations and identifies remaining gaps and critical barriers to progress in this area.

BIG EARTH DATA IN SUPPORT OF THE SUSTAINABLE DEVELOPMENT GOALS

Dr. Chunlin Huang, a researcher with the Northwest Institute of Eco-Environment and Resources, Chinese Academy of Sciences, spoke on behalf of the scheduled speaker Zhongchang Sun, associate researcher with the Institute of Remote Sensing and Digital Earth, Chinese Academy of Sciences. Dr. Huang highlighted the role of big Earth data for implementing the Sustainable Development Goals (SDGs). He began by describing world urbanization between 1970 and 2030; there were only three megacities in 1970, but in 2030, the number of megacities will exceed 40 (see Figure 5-1). While SDG 11 in the 2030 Agenda aims to make cities and human settlements inclusive, safe, resilient, and sustainable, there is a need for additional data to monitor and assess the progress of SDG indicators, Dr. Huang stated.

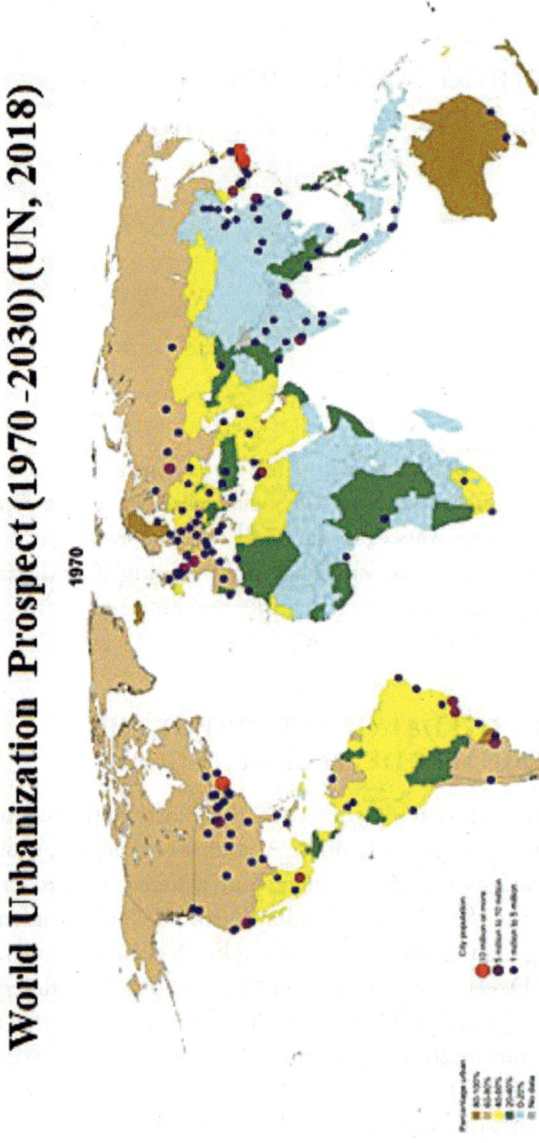

FIGURE 5-1 World urbanization prospect between 1970 and 2030.
SOURCE: Chunlin Huang, Presentation, National Academies of Sciences, Engineering, and Medicine, December 16, 2019, Washington, DC, based on United Nations, Department of Economic and Social Affairs, Population Division. 2018. World Urbanization Prospects 2018. Available at https://population.un.org/wup/Maps. Accessed April 16, 2020. Licensed under Creative Commons license CC BY 3.0 IGO.

Dr. Huang described efforts by the Chinese Academy of Sciences' Big Earth Data Science Engineering (CASEarth) program, whose mission is to build the world's leading big Earth data-sharing platform to monitor and assess selected indicators relating to SDGs 2 (zero hunger), 6 (clean water and sanitation), 11 (sustainable cities and communities), 14 (life below water), and 15 (life on land). Figure 5-2 provides information about the CASEarth structure and work.

One CASEarth project, said Dr. Huang, is examining the proportion of the population with easy access to public transportation in China. The project is using computational methods that extract data related to public transport (transit, subway) and calculate the proportion of the population that has convenient access within the urban built-up area, among other data. The analysis indicated that the proportion of the population with convenient access to public transportation at the provincial level was 64.28 percent on average. Also, at the prefecture-level city scale, the proportion of the population having convenient access to public transportation in densely populated cities was generally higher than in sparsely populated cities.

Regarding decision support, Dr. Huang noted that their work is providing support for the comprehensive evaluation of sustainable development in Chinese cities. The report *Big Earth Data in Support of the Sustainable*

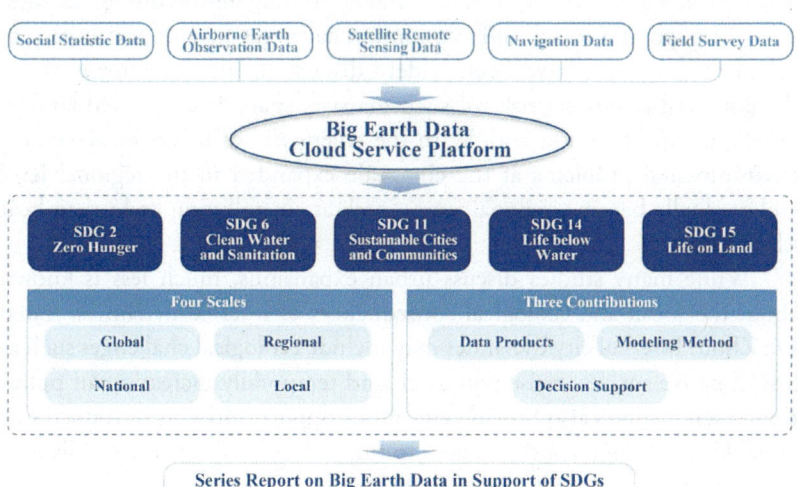

FIGURE 5-2 CASEarth Big Earth Data Cloud Service Platform.
SOURCE: Chunlin Huang, Presentation, National Academies of Sciences, Engineering, and Medicine, December 16, 2019, Washington, DC.

Development Goals in 2019,[1] which was recently released, is an important contribution toward the 2030 Agenda for Sustainable Development. The report includes 12 case studies centered on the 11 SDG indicators within five selected goals and covers the perspectives of data, methods and models, and decision support at local, national, regional, and global scales.

DATA GAPS AND URBAN EXPANSION IN CHINA

Weiqi Zhou, professor of urban ecology at the Research Center for Eco-Environment Studies, Chinese Academy of Sciences, emphasized that urbanization in China has large spatial variations, still mostly occurring in certain regions on the east coast. The concentration of urban expansion in certain regions has resulted in the emergence of urban megaregions. Existing cities, expanding suburbs, and new urban settlements and infrastructures have been gradually woven into urban megaregions. These megaregions concentrate on population and economic activities.

According to the National New-Type Urbanization Plan, China's first official plan on urbanization released in 2014, the urban megaregion or urban agglomeration will be the main type of urban spatial form over the next decade, said Dr. Zhou. China has planned to invest heavily in facilitating the formation and growth of urban megaregions. More unplanned urban megaregions will be emerging and expanding in the coming decades.

Although the social and economic challenges and opportunities of urban megaregions have been widely discussed, little is known about the potential ecological risks of such massive, spatially connected land of development, Dr. Zhou said. With the emergence of urban megaregions, environmental problems at the city scale expanded to the regional level and gradually became regional issues, such as air pollution and urban heat island effect.

While many studies discuss urban expansions, much less is known about the social and ecological consequences of internal dynamics, stated Dr. Zhou. Internal city dynamics may include ecological challenges such as loss of native species, noise pollution, and temporally increasing air pollution. Such changes also provide enormous opportunities to introduce sustainable technologies and practices to make the city greener, more livable, resilient, and energy and resource efficient.

---

[1] Chinese Academy of Sciences. 2019. Big Earth Data in Support of the Sustainable Development Goals. Available at https://www.fmprc.gov.cn/mfa_eng/topics_665678/2030kcxfzyc/P020190924800116340503.pdf. Accessed March 9, 2020.

Dr. Zhou discussed data gaps, noting that there is currently an overabundance of data, but much of the data has not been converted to information that can inform policy change. For example, there is extensive Earth observation data that has been archived and collected daily; however, one of the challenges is how to convert these data to inform decision making. There are significant amounts of biophysical data, but a lack of socialeconomic data, especially at the finer level. In some places where the population exceeds 1 million, only one indicator exists to gauge socioeconomics.

Dr. Zhou said that policy makers need more actionable knowledge that is immediately relevant and easy to use to inform science-based policies, particularly at the local scale. There is a need to move beyond transdisciplinary knowledge to co-production by working together with managers and decision makers to co-develop new science and tools for urban sustainability (see Figure 5-3).

LESSONS FROM THE EARTH FROM SPACE INSTITUTE'S INAUGURAL SYMPOSIUM

Miguel Román, founding director of the Earth from Space Institute (EfSI), an independent program of the Universities Space Research Association (USRA), noted that many of the primary challenges in managing rapid urbanization—from ensuring adequate housing and infrastructure to support growing populations, to confronting the environmental impact of urban sprawl, to reducing vulnerability to disasters—are exacerbated by a dearth of information tailored to the challenges and needs of cities. For decades, urban researchers in the United States have had little input in driving the programmatic priorities of earth science research on sustainability issues. For instance, the 2017–2027 Decadal Survey for Earth Science and Applications from Space[2] placed little emphasis on how to consider the sustainability of urban areas, concentrating instead on the traditional study of urbanization as a land transition. As a result, future satellite observations will continue to monitor urban areas merely as a type of land cover, using space-based data and technology requirements adapted from other disciplines (e.g., agriculture and forestry). To thrive under a changing planet, according to Dr. Román, the earth science community must, therefore, acknowledge the unique data and information needs of

---

[2] See https://sites.nationalacademies.org/DEPS/ESAS2017/index.htm. Accessed March 9, 2020.

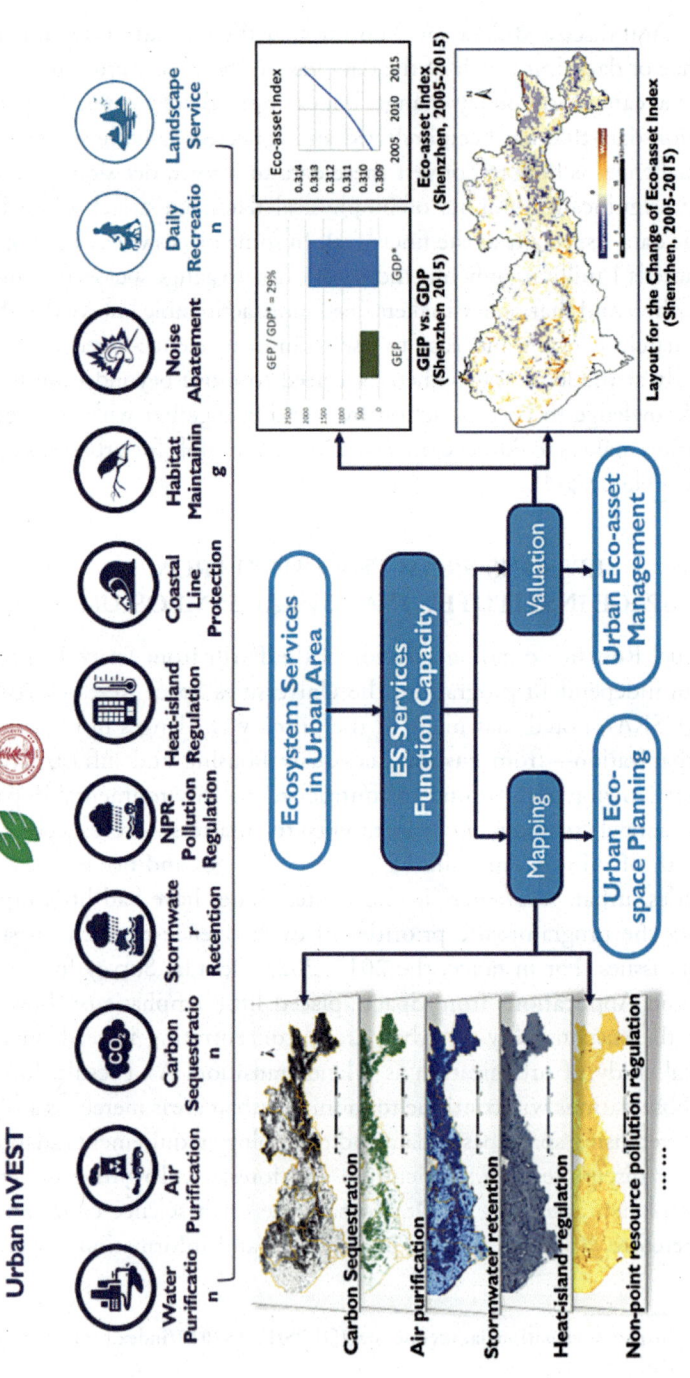

FIGURE 5-3 Developing new tools to support urban planning and management toward co-production of knowledge.
SOURCE: Weiqi Zhou, Presentation, National Academies of Sciences, Engineering, and Medicine, December 16, 2019, Washington, DC.

cities and consider alternative solutions that drive scientific progress toward urban sustainability.

Dr. Román also discussed that a second challenge for the earth science community is providing timely data to settlements on disaster risk reduction. Three sustainable development goals are related to disaster risk reduction (SDG 9 on infrastructure, SDG 11 on cities, and SDG 13 on climate action). In disasters, earth science data have focused on describing the hazards themselves but have done little to help communities reduce their risk and bolster their resilience. Dr. Román commented that in the United States, disaster risk management is addressed in a reactive manner—"We wait for things to break and then we try to fix them," he said. He also discussed how, in disaster situations, the data available are not useful if not contextualized and packaged in a timely manner, especially during rapid and needs-based assessments.

A recent USRA EfSI symposium, *Making Communities More Resilient to Extreme Flooding* held on October 30–31, 2019, focused on issues related to disaster risk management with 160 attendees from 95 organizations.[3] Dr. Román reported some key themes of the symposium included: (1) discussing how to improve communication across sectors; (2) considering flooding as an integrated system, including nature-based solutions in the context of flood risk management; (3) using visualization as an aid to improve perceptions of flood risks; (4) finding the right data and applying them; and (5) strengthening community resilience through public–private partnerships. During a breakout session on flood risk management in rapidly urbanized areas, participants discussed that investing in space-based data and analytical tools that could inform local action was urgently needed, Dr. Román added.[4]

Dr. Román discussed how, although sustainability is a good target in theory, survivability is often the focus on the ground. While the United States and China have resources to allow decision makers to think about how to adapt cities to better deal with a weather disaster, poorer regions (e.g., Small Islands and Developing States) often do not have that luxury. Citizens are not thinking about long-term adaptation. They have to rely on

---

[3] See symposium program and summaries at: https://www.usra.edu/efsi-2019-symposium. Accessed March 9, 2020.

[4] Related reading includes: Stokes, E. C., and K. C. Seto. 2019. Characterizing urban infrastructural transitions for the Sustainable Development Goals using multi-temporal land, population, and nighttime light data. *Remote Sensing of Environment* 234:111430. https://doi.org/10.1016/j.rse.2019.111430.

informal systems of disaster response and aid both as a life-saving resource and as a community-building resource.

One takeaway from the October symposium, he related, was how important it is to bring the media to the table to learn how to communicate and improve public perceptions of risk. Dr. Román described how the media have begun to shift in discussion of disasters. For example, in the field of broadcast meteorology, the focus is often on describing a hazard-based assessment of hydrometeorological variables (e.g., rainfall, pressure, and wind speed), as well as short-term impacts. But increasingly, meteorologists are building a longer-term narrative about weather patterns: linking disasters to climate change, discussing the increasing frequency of events, tracking a disaster's sustained damage over time, and its distributional impacts. In journalism terms, he noted they are discussing "who" has been most affected, "what" is the capacity of the affected communities to cope, and "where" to send aid.

These kinds of questions are also increasingly included in scientific studies of disasters. For example, in Puerto Rico, scientists used satellite-derived estimates of outdoor illuminations to track the number of days without electricity following Hurricane Maria in 2017 (see Figure 5-4)[5] and to understand why some communities shouldered longer-term impacts than others. Dr. Román illustrated with the example of Puerto Rico that data on electricity restoration could help inform the U.S. Federal Emergency Management Agency and the Department of Housing and Urban Development about where to direct recovery funds to address the most underserved communities. These are the kind of relevant data that can be used to inform a needs assessment.

Nongovernmental organizations (NGOs) also need to improve their assessment and communication of distributional impacts from major disasters. The symposium's discussions highlighted that NGOs have a very small time window to seek out emergency financial assistance, Dr. Román said. As such, there is a need for near-real time data that can enable NGOs to better visualize and communicate the conditions of underserved communities on the ground. Dr. Román added that his organization is coordinating with entities that are distributing resources and life-saving supplies

---

[5] Román, M. O., E. C. Stokes, R. Shrestha, Z. Wang, L. Schultz, E. A. S. Carlo, Q. Sun, B. Jordan, A. Molthan, V. Kalb, C. Ji, K. C. Seto, S. N. McClain, and M. Enenkel. 2019. Satellite-based assessment of electricity restoration efforts in Puerto Rico after Hurricane Maria. *PLoS One* 14(6): e0218883. https://doi.org/10.1371/journal.pone.0218883. Accessed March 9, 2020.

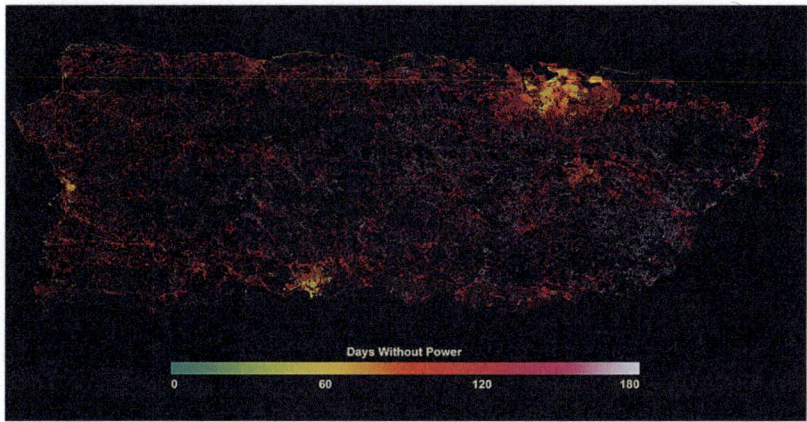

FIGURE 5-4 Satellite-based assessment of electricity restoration efforts in Puerto Rico after Hurricane Maria.
SOURCE: Miguel Román, Presentation, National Academies of Sciences, Engineering, and Medicine, December 16, 2019, Washington, DC.

(e.g., in the aftermath of Hurricane Dorian in the Bahamas).[6] They are also examining how to close data gaps through mapping, as well as how space-based data can be used to improve coverage, for example, in Haiti where there is little information on settlements, roads, and infrastructure mapped before an event.[7]

Dr. Román also discussed the status of the U.S. National Flood Insurance Program (NFIP),[8] which is up for congressional reauthorization. Traditionally, the program has incentivized individuals and developers to build in hazardous areas. He believes that ongoing policy discussions need to take place to revamp the NFIP and offer sustainable solutions that are ready to be scaled up and replicated across disaster-afflicted communities, both on the coast and inland.

---

[6] See https://www.washingtonpost.com/weather/2019/09/13/dorians-wrath-satellite-imagery-shows-northwestern-bahamas-dark-after-disastrous-storm. Accessed March 9, 2020.

[7] Goldblatt, R., N. Jones, and J. Mannix. 2020. Assessing OpenStreetMap completeness for management of natural disaster by means of remote sensing: A case study of three small island states (Haiti, Dominica and St. Lucia). *Remote Sensing* 12:118. https://doi.org/10.3390/rs12010118. Accessed March 9, 2020.

[8] See https://www.fema.gov/national-flood-insurance-program. Accessed March 9, 2020.

DISCUSSION

Participants discussed a number of issues related to the use of earth observations to support decision making. For example, related to post-disaster and recovery scenarios, there is a need to incorporate satellite-derived data to communicate the compounding risks of disasters effectively. Dr. Román noted the need for data to help build social capital internally, empathy externally, and then integrate the data in a way that can be communicated effectively. New satellite missions specifically designed to measure urbanization as a "process," instead of urbanization as a "place" (i.e., the Urban 2.0 data paradigm), will help stakeholders identify sectoral interlinkages and their implications across economic, social, and environmental areas in implementing the SDGs.

Several participants also discussed the need to communicate with stakeholders and particularly decision makers. Dr. Román added that the outcome of what to communicate needs to be at the right time and spatial scales to be relevant.

6

Addressing Key Intersecting Issues in Urban Sustainability

Throughout the workshop, speakers and participants discussed the importance of examining urban sustainability through interdisciplinary and multidisciplinary lenses. Workshop participants divided into three groups to discuss key issues related to the intersection of urban climate change mitigation and adaptation, urban health, and sustainable transportation, including green infrastructure and urban flooding in China and the United States. The discussion below summarizes some main points from each of the breakout groups.

CIRCULAR ECONOMY AND GREEN INFRASTRUCTURE

Participants in this breakout session discussed challenges and opportunities for a circular economy transition in cities with the support of energy innovation and green infrastructure, including those related to critical water and waste issues.

Common themes and issues discussed during the session, as highlighted by session moderator Yong-Guan Zhu, director general of the Institute of Urban Environment, Chinese Academy of Science, included:

- *Recognizing knowledge gaps around green infrastructure.* Participants discussed that the definition of green infrastructure is problematic. Constructing a framework to quantify the benefits of inputs and outputs and a methodology to evaluate the co-benefits of green infrastructure would strengthen research, several suggested. This would include understanding the benefits of initiatives related to green space, biodiversity conservation, and climate change, among other areas.

- *Assessing the impact of green infrastructure.* There is a gap in understanding of the effectiveness of green infrastructure as compared to traditional engineering solutions. Participants described the robustness of green infrastructure and the need to develop additional systems, for example, those that can harvest the green biomass for energy or for other uses.
- *Addressing e-waste.* Participants also discussed that green infrastructure solutions will not be able to address challenges related to e-waste directly. More research and technology will be needed to address this issue in the urban circular economy.
- *Addressing organic waste.* This waste can be utilized in urban/peri-urban food production and urban green infrastructure. For example, some material after treatment can be fed back into green infrastructure processes; however, more innovation is needed to make the system more robust and economically viable in the field.
- *Engineering solutions for urban systems.* Participants discussed the need for a better code of practices for green infrastructure, including more standardization. Standards of practice would help ensure the robustness of nature-based solutions and monitor effectiveness.
- *Decision making and urban sustainability.* There are knowledge gaps in terms of how institutional structures and divided decision-making power affect nature-based solutions (e.g., dividend power between municipalities).
- *Monitoring of nature-based solutions.* There is a need to improve what is known about urban monitoring, particularly in examining the impact of a nature-based solution.
- *Communication around urban issues.* Participants discussed methods for simplifying complex information on urban sustainability issues into ways to communicate with urban planners.

PEOPLE-CENTRIC DESIGN FOR SUSTAINABLE CITIES

Participants in this breakout session discussed various issues related to moving toward people-centric design for sustainable cities. Some topics included climate migrants and well-being, livability of cities and neighborhoods, provision of social service, preparation for the aged society, and science-policy action to connect economic development activities and job creation.

Moderator Frances Colón, chief executive officer of Jasperi Consulting, summarized the key points made by participants during the discussion, including:

- *The role of citizens.* How citizens can inform critical issues in urban sustainability was a key topic discussed, including identifying mechanisms to better integrate citizens into decision-making.
- *Consideration of inequities in urban design.* The first step for scientists in urban planning is to listen to poor, low-resource communities. Participants suggested that urban planning must be done with the most vulnerable in mind, first and foremost.
- *Moving beyond greenhouse gases (GHGs).* Urban sustainability efforts cannot focus on GHG emissions solely, they must also examine, for example, adaptation to climate change impacts, vulnerability of front-line communities, and densification of urban cities that can create heat islands, when a metropolitan area is warmer than the surrounding rural areas.
- *Broadening participation.* Understanding who is and should be at the table (e.g., community members, scientists, decision makers, business sector) to more fully engage on urban sustainability issues is critical.
- *Incorporating the pedestrian into design.* People experience the city on the street level, and that must be considered when planning a city from above. There is a nuance to the pedestrian experience that cannot be found through satellite data.
- *Role of decision makers.* Decision makers' role in these issues, including the shortness of some political cycles and who makes decisions in the government, was discussed. The importance of the political process and the impact it has on these decisions must not be underestimated.
- *Communicating data to inform decisions.* Science communication and how communities are educated about these issues was also a key issue discussed. If data are not conveyed to decision makers in a way that matters to them, the data will not create change.

BENEFITS OF CROSS-SECTORAL APPROACHES FOR URBAN SUSTAINABILITY

Participants during this breakout session discussed the water-food-health nexus, synergies between environmental management and social development, and linkages between air quality conservation and green public transportation. Moderator Chengri Ding, professor in the School of Architecture, Planning, and Preservation at University of Maryland, summarized key topics discussed during the session:

- *Benefits of cross-sectoral approaches.* Regarding benefits of cross-sectoral approaches for urban sustainability, the group discussed the need to:
  o Allow researchers to examine urban areas as a whole, including water-food-health-transportation issues;
  o Allow researchers to identify areas they are struggling with, research gaps, and contexts affecting process and outcomes for urban sustainability, such as geographic scales, role of the local context, and cultural content;
  o Enable several scientific disciplines to communicate with each other;
  o Understand differences and similarities, including how to translate one solution into other places and build up urban science;
  o Conduct a systematic review on what works, what does not work, and over what scale, including issues related to transportation (actual vehicles), energy-sector transformation (renewables), and manufacturing sectors, etc.
- *Barriers to cross-sectoral collaboration.* These barriers related to urban sustainability include the lack of collaboration among different disciplines and a need for education and next generation training through an interdisciplinary approach. Participants agreed that one mechanism that works for one city in the promotion of sustainable growth may not work for other cities or the city as a whole. There is a need to look at specific cases.
- *Developing and broadening partnerships.* In terms of mechanisms to develop effective partnerships to support cross-sectoral approaches, the group discussed the Intergovernmental Panel on Climate Change (IPCC) report and Sustainable Development Goals. These

efforts address not only urban issues, but also integrating other relevant challenges such as transportation, food, and health.
- *Role of basic research.* There is also a need for basic research in order to solve problems with a full understanding of urban systems and communities.

7

The Way Forward: Future Needs and Opportunities

In the final session of the workshop, planning committee chair Karen Seto led a discussion of possible future steps and opportunities based on the workshop discussions. She encouraged participants to consider tangible actions and what issues the National Academies could amplify, among others. Given the differences between the United States and China, the efforts can be complementary, she added.

Dr. Seto summarized the discussion from the workshop, including several cross-cutting themes she heard:

Generational shift toward action. One issue discussed during several presentations was the generational shift that has been occurring around urban sustainability: that is, a demand for action by young people on these issues, particularly climate change. Several speakers noted that behavior change may happen more quickly with this generational shift, she noted. Students at the university level are critical to this movement, and several participants urged a consideration of how to harness this energy.

Interdisciplinary and multidisciplinary approaches to urban sustainability. Several speakers discussed the need for interdisciplinary approaches for addressing urban challenges, especially the value of bringing a different allied lens to the table. There is a need to be creative in bringing in other disciplines to face sustainability challenges, for example, the humanities, design and art, and psychology. While it is very clear that more data are not going to lead to better decisions, there are new ways to broaden understanding of these issues, such as by examining happiness, emotions, or empathy. This is where the allied fields or issues can contribute in thinking about urban sustainability, said Dr. Seto.

Spatial and temporal scales matter. Several speakers discussed the critical importance of spatial and temporal scales in urban sustainability challenges; in fact, several speakers noted that the wrong answers may result if the same spatial or temporal scales are assumed. It is clear that context matters and what is applicable across different places will differ.

Partnerships and practice. The need for stronger partnerships to support urban sustainability research and practice was discussed continuously during the workshop, Dr. Seto observed. Universities have a dual mission to research and teach, said Dr. Seto, but what is new in terms of thinking about urban sustainability is that the students also want to take action. It is not just that they want to learn theory, she continued, they want to learn how to implement it. She asked how to make them citizen-scientists and amplify their impact, and she suggested that bridging practice and knowledge in new ways and collaborating outside of university settings will drive this change.

Locked-in behavior. Several speakers discussed the role of locked-in behavior in understanding urban sustainability issues, such as related to car usage. Once behaviors are locked in, they are very entrenched, she observed.

Need for nature-based solutions. Several workshop participants discussed the need to train a workforce that can design, implement, and maintain nature-based solutions, said Dr. Seto. Participants discussed the need to develop codes and legal systems to protect nature-based solutions. This is a major knowledge gap as the focus has been almost primarily on more traditional engineering solutions, Dr. Seto stated.

Experimentation and innovation. Several participants discussed experimentation, such as innovation around net zero neighborhoods and greenscapes. More information about how to scale these efforts up is needed. Scaling up and partnerships is a big question for universities, the Academies, and others.

Policy design matters. Several speakers discussed the topic of how policy design matters in urban sustainability, Dr. Seto said. Linked to that is the critical importance of public education and communication.

Changes in public education and communication around urban sustainability. Given that most of the world now lives in urban areas, there is a need to rethink education and communication around urban sustainability in a way that can more fully engage potential students and the public. Students, at least in the United States, learn a lot about how the national government works; however, most children probably do not learn about the geography of their local city and may have no idea how their local cities are run. There is a big shift that needs to take place in thinking about urban sustainability and training the next generation, stated Dr. Seto, not only at the university level but also for K-12 students.

Role of citizens. Workshop participants discussed the importance of engaging citizens in urban sustainability issues. The key question is how to engage the average citizen to be interested in these issues.

Appendix A

Workshop Agenda

**Advancing Urban Sustainability in China and the United States
A National Academies Workshop in Collaboration with
the Chinese Academy of Sciences**

December 16, 2019

National Academy of Sciences
Lecture Room
2101 Constitution Ave NW
Washington, DC

9:00 am   **Welcome from the National Academy of Sciences**
                 Marcia McNutt, President, U.S. National Academy of Sciences

9:10 am   **Opening Remarks from the Chinese Academy of Sciences**
                 Yaping Zhang, Vice President, Chinese Academy of Sciences

9:20 am   **Overview and Goals of the Workshop**
                 Karen Seto (NAS), Yale University, Committee Chair
                 Franklin Carrero-Martínez, National Academies of Sciences,
                    Engineering, and Medicine

9:30 am   **Framing Remarks: Current Landscape for Sustainable Urbanization Research and Practice in China and the United States**
- Deb Niemeier (NAE), University of Maryland (U.S. Perspective)
- Wei-Qiang Chen, Institute of Urban Environment, Chinese Academy of Sciences (Chinese Perspective)

10:00 am   Q&A and Discussion

10:15 am   **National Science Foundation's Interaction with China on Urban Sustainability Research**
Linda Blevins, National Science Foundation

10:30 am   BREAK

10:45 am   **Panel I: Urban Sustainability Research Activities at a University Level**
Moderator: Susan Hanson (NAS), Clark University
- Jianming Cai, Institute of Geographical Sciences and Natural Resources Research, Chinese Academy of Sciences
- Luis Bettencourt, University of Chicago
- Yan Song, University of North Carolina

11:30 am   Q&A and Discussion

11:45 am   LUNCH

12:45 pm   **Panel II: Architecture, Urban Design, and Sustainable Cities in China and the U.S.**
Moderator: Robert Cervero, University of California, Berkeley
- Kongjian Yu, Peking University
- V. Kelly Turner, University of California, Los Angeles
- Yingling Fan, University of Minnesota
- Matei Georgescu, Arizona State University

1:45 pm   Q&A and Discussion

2:00 pm	**Panel III: Data and Earth Observation for Decision Making**
	Moderator: Jiahua Pan, Chinese Academy of Social Sciences
	• Chunlin Huang, Northwest Institute of Eco-Environment and Resources, Chinese Academy of Sciences, speaking for Zhongchang Sun, Institute of Remote Sensing and Digital Earth, Chinese Academy of Sciences
	• Weiqi Zhou, Research Center for Eco-Environment Studies, Chinese Academy of Sciences
	• Miguel Román, Universities Space Research Association
2:45 pm	Q&A and Discussion
3:00 pm	BREAK and Preparation for Breakout Sessions
3:20 pm	**Breakout Sessions**
	1. **Circular Economy and Green Infrastructure** (Lecture Room) Moderator: Yong-Guan Zhu, Institute of Urban Environment, Chinese Academy of Science
	2. **People-Centric Design for Sustainable Cities** (Board Room) Moderator: Frances Colón, Jasperi Consulting
	3. **Benefits of Cross-Sectoral Approaches for Urban Sustainability** (Room 125) Moderator: Chengri Ding, University of Maryland
4:20 pm	**Report Back from Breakout Groups** Session Moderators
4:35 pm	**The Way Forward: Future Needs and Opportunities** All Participants with Karen Seto (NAS), Yale University, Moderator
5:00 pm	**Summary Remarks** Karen Seto (NAS), Yale University, Committee Chair
5:15 pm	**Adjourn Workshop**

Appendix B

Biographies of Speakers and Moderators

KAREN SETO (NAS) (Planning Committee Chair) is the Frederick C. Hixon Professor of Geography and Urbanization Science at Yale University. An urban and land change scientist, she is one of the world's leading experts on contemporary urbanization and global change. Her research focus is on how urbanization will affect the planet. She has pioneered methods to reconstruct urban land use with satellite imagery and has developed novel methods to forecast urban expansion. She has conducted urbanization research in China for 20 years and in India for more than 10. Her research has generated insights on the links between urbanization and land use, food systems, biodiversity, and climate change. Dr. Seto has served on numerous national and international scientific bodies. She is co-leading the urban mitigation chapter for the Intergovernmental Panel on Climate Change (IPCC) 6th Assessment Report and co-led the same chapter for the IPCC 5th Assessment Report. She is co-editor-in-chief of the journal *Global Environmental Change*. From 2000 to 2008, she was faculty at Stanford, where she held joint appointments in the Woods Institute for the Environment and the School of Earth Sciences. She has received numerous awards for her scientific contributions, including the Outstanding Contributions to Remote Sensing Research Award from the American Association of Geographers. Dr. Seto is an elected member of the U.S. National Academy of Sciences, the Connecticut Academy of Science and Engineering, and the American Association for the Advancement of Science. She earned a Ph.D. in geography from Boston University.

LUIS BETTENCOURT is the inaugural director of the Mansueto Institute for Urban Innovation and professor of ecology and evolution at the University of Chicago, as well as an external professor of complex systems at the Santa Fe Institute. He has held postdoctoral positions at the University of Heidelberg (Germany), Los Alamos National Laboratory (Director's Fellow and Slansky Fellow), and Massachusetts Institute of Technology (Center for Theoretical Physics). He has worked extensively on complex systems theory and on cities and urbanization, in particular. His research emphasizes the creation of new interdisciplinary synthesis to describe cities in quantitative and predictive ways, informed by classical theory from various disciplines and the growing availability of empirical data worldwide. He is the author of over 100 scientific papers and several edited books. His research has been featured in leading media venues, including the *New York Times*, *Nature*, *Wired*, *New Scientist*, and *Smithsonian*. He was trained as a theoretical physicist and obtained his undergraduate degree from Instituto Superior Técnico (Lisbon, Portugal) and his Ph.D. from Imperial College (University of London, UK) for research in statistical and high-energy physics models of the early Universe.

LINDA BLEVINS joined the National Science Foundation (NSF) as the deputy assistant director of the Engineering Directorate in December 2017. For more than a decade prior to joining NSF, she was a senior technical advisor in the Office of the Deputy Director for Science Programs in the U.S. Department of Energy's Office of Science, where she provided technical and policy advice on all aspects of science funding program management. She previously served as an NSF program director, as a senior member of the technical staff at Sandia National Laboratories, and as a National Institute of Standards and Technology research staff member. Her research expertise is in combustion. Dr. Blevins received a Ph.D. from Purdue University, an M.S. from Virginia Tech, and a B.S. from the University of Alabama, all in mechanical engineering.

JIANMING CAI is a full professor at the Institute of Geographical Sciences and Natural Resources Research (IGSNRR) of the Chinese Academy of Sciences (CAS), and the director of RUAF (Resource Centres on Urban Agriculture and Food Security) China, affiliated to the RUAF Foundation based in the Netherlands. He has published more than 160 papers in academic journals both in Chinese and English, plus more than 100 key consultant reports. Dr. Cai frequently serves as a senior consultant or expert on sustainable urbanization, regional and urban development, regional cooperation and integration, culture-embedded place-making, urban agriculture and food security to international agencies, such as the United Nations Human Settlements Programme (UN-HABITAT), Food and Agriculture Organization of the United Nations (FAO), United Nations Industrial Development Organization (UNIDO), World Bank, European Union, Asian Development Bank, Ford Foundation, International Development Research Centre (IDRC), German Technical Cooperation Agency (GTZ), Directorate-General for International Cooperation (DGIS), and the Lincoln Institute. He has also advised various Chinese government agencies, from the central to local levels, as well as private-sector companies such as Shui On Land. He is an active member of many academic associations including the vice-chairman of the Chinese Urban Agriculture and Recreational Agriculture Association as well as a board member of the Association of Administration Regionalization, Chinese Geographic Society Rural Development Division, and Urban and Regional Planning Association. His current research focuses on urban-rural sustainable development with emphases on urban and peri-urban agrotourism, urban-rural linkages, spatial restructuring from cultural perspectives, urban renewal, and innovative space making. Dr. Cai received his first degree in urban planning and economic geography from Beijing University, his master's in geographical information systems for urban applications from ITC of the Netherlands, and his Ph.D. in sustainable urban development from the University of Hong Kong.

FRANKLIN CARRERO-MARTÍNEZ joined the National Academies of Sciences, Engineering, and Medicine in 2018, where he directs the Global Policy and Development and the Science and Technology for Sustainability program within the Division of Policy and Global Affairs. Prior to this appointment, he was the acting deputy science and technology adviser to the U.S. secretary of state. His multidisciplinary career includes several roles in academia and government: from researcher and educator, science administrator, to science policy and diplomacy. Previously, Dr. Carrero-Martínez held appointments as an associate professor at the University of Puerto Rico, Mayagüez, adjunct professor at the UPR Medical Science Campus, and visiting scholar at Duke University, Massachusetts Institute of Technology, and Japan's Institute of Genetics. Dr. Carrero-Martínez started his career in science diplomacy and policy as the American Association for the Advancement of Science's Roger Revelle Fellow in Global Stewardship. He served this prestigious fellowship with a joint appointment between the Office of the Science and Technology Adviser to the Secretary of State (STAS) and the National Academy of Sciences. At the end of his fellowship, he served as program director at the National Science Foundation supporting the foundation's diplomatic and representational obligations, while managing a portfolio of international basic science collaboration grants before returning to STAS in 2016. As the Department's senior advisor on science, technology, and innovation issues (STI), he directed the STAS Office. In this role he provided senior officials with analysis, guidance, recommendations, and strategic planning to anticipate the foreign policy impacts of emerging STI issues, built STI capacity within the Department, and engaged the National Security Innovation Base to promote Department priorities. Dr. Carrero-Martínez holds a B.S. in biology with honors from the University of Puerto Rico, a Ph.D. in cell and developmental neurobiology, and a certificate in business administration from the University of Illinois at Urbana-Champaign.

ROBERT CERVERO (Planning Committee Member) is professor emeritus of city and regional planning at the University of California, Berkeley, where he chaired the department and directed the Institute of Urban and Regional Development and the University of California Transportation Center. His research focuses on sustainable transportation policy and the nexus between urban transportation and land-use systems. He has written numerous books and articles in these areas, including *Beyond Mobility*, which won the 2019 National Urban Design Best Book Award. Currently, he is working with AECOM on a study of transit joint development for the U.S. Transit Cooperative Research Program, is a faculty affiliate at NYU-Abu Dhabi and Tongji University in Shanghai, and serves on the international advisory panel for Dubai's 2040 strategic master plan. He was a contributing author to the IPCC Fifth Assessment and UN-HABITAT's Global Report on Sustainable Mobility. He recently received the Athena Accolade from KTH University, the Distinguished Legacy Award from ITS at the University of California, Berkeley, and was recognized by *Planning* magazine as one of the top planning academics in the United States based on Google Scholar h-index citations, naming him "the world's top expert on transit-oriented development." Dr. Cervero received his Ph.D. in urban planning from the University of California, Los Angeles.

WEI-QIANG CHEN is a professor of resources and urban sustainability science at the Institute of Urban Environment, CAS. He received his undergraduate and Ph.D. degrees from the School of Environment at Tsinghua University, Beijing, and was at Yale School of Forestry and Environmental Studies from 2010 to 2015. His research focuses on (1) urban metabolism and urban complexity, and (2) anthropogenic cycles and trade of resources, especially metals and plastics. Dr. Chen's studies have been published in the *Proceedings of the National Academy of Sciences of the United States of America*, *Environmental Science and Technology*, and other first-level journals. He was elected to the board of the International Society for Industrial Ecology in 2018, and was the founding president of the Chinese Society for Industrial Ecology created in 2015. He is now associate editor for the journals *Resources, Conservation, and Recycling* and *Journal of Industrial Ecology*.

FRANCES COLÓN (Planning Committee Member) is chief executive officer of Jasperi Consulting and a 2019 recipient of the Leadership in Government Fellowship of Open Society Foundations, an initiative supporting seasoned public servants chosen from the senior ranks of federal, state, and local government who have advanced economic and social change. Dr. Colón is the former deputy science and technology adviser to the Secretary of the U.S. Department of State (2012–2017) where she promoted the integration of science and technology into foreign policy priorities. In her role as science diplomat, Dr. Colón led reengagement of scientific collaboration with Cuba and coordinated climate change policy for the Energy and Climate Partnership of the Americas announced by President Obama. Dr. Colón is founding board member of Cenadores Puerto Rico, a nonprofit that provides a platform for the Puerto Rican diaspora and friends of Puerto Rico to strengthen civil society on the island. Dr. Colón is a graduate of the National Hispana Leadership Institute, and fellow of the U.S.-Japan Leadership Program, the National Committee on U.S.-China Relations, and the Austria Leaders Program. In 2016, Dr. Colón was named one of the 20 most influential Latinos in technology by CNET. She is the recipient of the Hispanic Heritage Foundation's 2015 Inspira Award and a 2015–2016 Google Science Fair judge. Dr. Colón earned her Ph.D. in neuroscience in 2004 from Brandeis University and her B.S. in biology in 1997 from the University of Puerto Rico.

CHENGRI DING is professor in the School of Architecture, Planning, and Preservation at the University of Maryland. Dr. Ding specializes in urban economics, urban and land policies, urban planning and policy, and China studies. He has published articles in leading journals such as *Journal of Urban Economics, Journal of Regional Science, Urban Studies, Environment and Planning B, Housing Policy Debates,* and *Land Use Policy.* Dr. Ding has written three books, five manuscripts in Chinese, and numerous publications in *China Journal.* He has been a principal investigator (PI) for many international policy projects on China, ranging from urban master planning, farmland protection, property tax reform, and local public financing. Dr. Ding has given over 50 invited or keynote speeches/presentations. He has been a consultant to the World Bank, Global Business Network, FAO, and leading Chinese agencies such as the National Development and Reform Commission (NDRC). He serves on the Advisory Board for the International Institute of Property Taxation. Dr. Ding is the founding director for the Lincoln Institute of Land Policy's China Program.

YINGLING FAN is a professor in the regional planning and policy area at the University of Minnesota and works interdisciplinarily in the fields of land use, transportation, social equity, and public health. Her overarching research goal is to investigate the impacts of spatial planning (e.g., land use, growth management, and transit improvements) on human activities and movements, as well as to understand the health and social aspects of such impacts. Her research combines ecological and behavioral analyses, most quantitatively, as a means of addressing urban sustainability challenges. Dr. Fan has published her work in various urban planning and transportation research journals. Her recent projects include investigating the impact of urban form on health disparities, the role of neighborhood and family structure in influencing leisure-time activity patterns, and the impact of transit corridor improvements on job accessibility and neighborhood change. She held the title of McKnight Land-Grant Professor from 2012 to 2014—a special award that recognizes and honors the University of Minnesota's most promising junior faculty. Dr. Fan has received the Collaborator of the Year Award from the Hennepin County-University Partnership, the Scholar Award from the Children, Youth and Family Consortium at the University of Minnesota, the Best Paper Award from the Transportation Research Board (TRB) Pedestrian Committee, and the TRB Patricia F. Waller Award. She holds a Ph.D. in city and regional planning from the University of North Carolina at Chapel Hill and a bachelor's degree in transportation engineering from Southeast University, Nanjing, China.

MATEI GEORGESCU is associate professor and associate director of the School of Geographical Sciences and Urban Planning in the College of Liberal Arts and Sciences at Arizona State University (ASU). He is also senior sustainability scientist and a member of the affiliated faculty for the Center for Biodiversity Outcomes at ASU's Julie Ann Wrigley Global Institute of Sustainability. Dr. Georgescu's research aims to improve understanding and characterization of distinct phenomena related to urbanization-induced landscape change. He focuses on identifying hydro-climatic and air quality impacts resulting from large-scale urbanization, as well as potential adaptation and mitigation strategies. In addition, Dr. Georgescu addresses environmental consequences (e.g., on climate and hydrology) of renewable energy expansion by integrating across physical, agricultural, and socioeconomic elements. The range of tools used to investigate these topics include climate models, remote sensing data and associated applications, and in situ weather/climate observations. Prior to joining ASU, Dr. Georgescu was a postdoctoral scholar in the Center on Food Security and the Environment (Department of Environmental Earth System Science) at Stanford University from 2008–2010. He received his Ph.D. in atmospheric science, M.S. in environmental sciences, and B.S. in meteorology from Rutgers University.

SUSAN HANSON (NAS) (Planning Committee Member) is a Distinguished University Professor Emerita and longtime professor of geography at Clark University. She is an urban geographer with interests in gender and economy, transportation, local labor markets, and sustainability. Her research has examined the relationship between the urban built environment and people's everyday mobility within cities; within this context, questions of access to opportunity, and how gender affects access, have been paramount. Her books include *Ten Geographic Ideas that Changed the World; Gender, Work, and Space* (with Geraldine Pratt); and *The Geography of Urban Transportation*. Dr. Hanson has been the editor of several academic journals, including *The Annals of the Association of American Geographers, Urban Geography*, and *Economic Geography* and has been the geography editor of the *International Encyclopedia of the Social and Behavioral Sciences*, first and second editions. She has led the School of Geography at Clark and is a past president of the Association of American Geographers (AAG), a fellow of the American Association for the Advancement of Science, a former Guggenheim Fellow, a former fellow at the Center for Advanced Study in the Social and Behavioral Sciences at Stanford, and a recipient of the Honors Award and of the Lifetime Achievement Award from the AAG and of the Van Cleef Medal from the American Geographic Society. In 2000 she was elected to the National Academy of Sciences and the American Academy of Arts and Sciences. She was recently division chair of the TRB of the National Research Council (NRC) and is on the TRB Executive Committee, is on the advisory board of the NRC's Gulf Research Program, and is on the editorial board of the *Proceedings of the National Academy of Sciences*. Her B.A. is from Middlebury College, and, before earning her M.S. and Ph.D. at Northwestern University, she was a Peace Corps Volunteer in Kenya.

CHUNLIN HUANG is a researcher with the Northwest Institute of Eco-Environment and Resources (NIEER) at CAS. From 2009 to 2010, he was a postdoctoral researcher with the Department of Civil and Environmental Engineering at the University of California, Los Angeles. He has published more than 50 papers in the Science Citation Index (SCI). Dr. Huang's current research activities concern hydrological remote sensing, hydrological data assimilation, and big earth data for monitoring Sustainable Development Goal (SDG) indicators. He received his Ph.D. in cartography and geographic information systems from the Cold and Arid Regions Environmental and Engineering Research Institute (CAREERI), CAS.

MARCIA McNUTT (NAS) is a geophysicist and the 22nd president of the National Academy of Sciences. From 2013 to 2016, she was editor-in-chief of *Science* journals. Dr. McNutt was director of the U.S. Geological Survey (USGS) from 2009 to 2013, during which time USGS responded to a number of major disasters, including the Deepwater Horizon oil spill. For her work to help contain that spill, Dr. McNutt was awarded the U.S. Coast Guard's Meritorious Service Medal. She is a fellow of the American Geophysical Union (AGU), Geological Society of America, American Association for the Advancement of Science, and International Association of Geodesy. Dr. McNutt is a member of the American Philosophical Society and the American Academy of Arts and Sciences, and a Foreign Member of the Royal Society, UK, and the Russian Academy of Sciences. In 1998, Dr. McNutt was awarded the AGU's Macelwane Medal for research accomplishments by a young scientist, and she received the Maurice Ewing Medal in 2007 for her contributions to deep-sea exploration. Dr. McNutt received a B.A. in physics from Colorado College and Ph.D. in earth sciences from the Scripps Institution of Oceanography.

DEB NIEMEIER (NAE) is the Clark Distinguished Chair of the Department of Civil and Environmental Engineering at the University of Maryland. Her primary research focus has been on developing highly accurate and accessible processes for emissions modeling and travel behavior models that can be used in the public sector, particularly the identification and modeling of environmental health disparities and those leading to improved understanding of formal and informal governance processes in urban and transportation planning. She is interested in emergent properties or characteristics that give rise to inequitable outcomes, particularly those associated with climate change. Prior to her appointment at the University of Maryland, she was a professor in the Department of Civil and Environmental Engineering and professor in the School of Education and Biological and Agricultural Engineering at the University of California, Davis. Dr. Niemeier became a fellow of the American Association for the Advancement of Science (AAAS) in 2014 and currently serves as past-chair of its Engineering Section. She was named a Guggenheim Fellow in 2015. In 2017, she was elected to the National Academy of Engineering (NAE) and recently completed service as a member of its Board on Energy and Environmental Systems. She served as editor-in-chief for *Transportation Research, Part A*, the leading international journal focused on transportation policy and practice, and was the first woman in the journal's history to hold this position. She received her Ph.D. in civil and environmental engineering from the University of Washington.

JIAHUA PAN (Planning Committee Member) is the director of the Institute for Urban and Environmental Studies at the Chinese Academy of Social Sciences (CASS). He is also a professor of economics at CASS Graduate School. Areas of interest include the economics of sustainable development, energy and climate policy, urban transformation, world economy, and environmental and natural resource economics. Dr. Pan worked for the United Nations Development Programme (UNDP) Beijing Office as an advisor on environment and development. He was also a lead author of the IPCC Working Group III 3rd, 4th, and 5th Assessment Report on Mitigation. He has been a member of the China National Expert Panel on Climate Change, a member of the National Foreign Policy Advisory Group and an advisor to the Ministry of Environment Protection. He was vice president of the Chinese Association of Urban Economy, Chinese Society of Ecological Economists, and Chinese Energy Association. Dr. Pan has been editor-in-chief of the *Chinese Journal of Urban & Environmental*

Studies and co-editor of *Climate Change 2001: Mitigation*. He has authored over 300 papers, articles and books in English (including in *Science, Nature,* and *Oxford Review of Economic Policy*) and Chinese (including in *Journal of Economic Research* and *China Social Sciences*). He has been awarded first and second prize of best research works, Chinese Academy of Social Sciences (2003, 2005, 2013), and was winner of Sun Yefang Economic Sciences Prize, 2011, and China Green Person of the Year 2010/2011. Dr. Pan received his Ph.D. at Cambridge University in 1992.

MIGUEL ROMÁN is the founding director of the Earth from Space Institute (EfSI), an independent program of Universities Space Research Association (USRA) dedicated to supporting the development of long-term strategies for reducing disaster risk and promoting community resilience, using the unique vantage point of Space. Dr. Román currently serves as the National Aeronautics and Space Administration's (NASA's) Terra, Aqua, and Suomi National Polar-orbiting Partnership's land discipline leader, helping manage a worldwide team of investigators in charge of generating long-term data records from the Moderate Resolution Imaging Spectroradiometer and the Visible Infrared Imaging Radiometer Suite, two of the largest and most comprehensive instrument suites ever launched to systematically monitor the planet's vital signs. Before joining USRA, Dr. Román served for 10 years as a civil servant scientist at NASA's Goddard Space Flight Center, where he pioneered the iconic Black Marble—a suite of satellite products that provide daily global views of Earth at night, with an emphasis on tracking the signatures of recovery across vulnerable communities affected by major disasters. Dr. Román has also led international activities under the Committee on Earth Observation Satellites and the Group on Earth Observations. President Barack Obama named him a recipient of the Presidential Early Career Award for Scientists and Engineers, the highest honor bestowed by the U.S. government on researchers beginning their independent careers. His writings have been featured in numerous news outlets, including NPR, *Washington Post,* NBC, *The Economist,* Telemundo, *Smithsonian Magazine,* and BBC World News.

YAN SONG is director of the Program on Chinese Cities and a professor in the Department of City and Regional Planning at the University of North Carolina at Chapel Hill. Dr. Song's research interests include low carbon and green cities, plan evaluation, land use development and regulations, spatial analysis of urban spatial structure and urban form, land use and transportation integration, and how to accommodate research in the above fields by using geographic information systems (GIS) and other computer-aided planning tools. Dr. Song's current research projects address domestic and international issues in the areas of impetus of urbanization and urban growth, tools of low carbon and green city developments, the efficacy of land and housing markets, effects of urban growth management regulations, and integration of urban land use and transportation plans. Her current research projects also document the evolution of China's urban land and housing policies and urban spatial structure in the era of China's transition toward a market economy. Dr. Song's research projects have been supported by the U.S. National Science Foundation, U.S. Environmental Protection Agency, U.S. Department of Housing and Urban Development, and Lincoln Institute of Land Policy. Dr. Song has served as a research affiliate at the National Center for Smart Growth at the University of Maryland and a faculty fellow at the Lincoln Institute of Land Policy. She has also served as a consultant on urban planning for the city government of Shenzhen, and a consultant on land use and transportation integration for the Beijing Municipal Institute of City Planning and Design in China.

ZHONGCHANG SUN is currently an associate researcher with the Institute of Remote Sensing and Digital Earth, Chinese Academy of Sciences. Dr. Sun received his Ph.D. in cartography and geographic information systems from the Centre for Earth Observation and Digital Earth, Chinese Academy of Sciences, Beijing, China, in 2011. He has authored or co-authored more than 20 SCI papers. His current research interests include urban environment remote sensing, urban sustainability, and land surface dynamics remote sensing.

V. KELLY TURNER is assistant professor of urban planning at the University of California, Los Angeles. Her research addresses the relationship between institutions, urban design, and the environment through two interrelated questions: How does urban design relate to ecosystem services in cities, and to what extent do social institutions have the capacity to deliver those services. Her approach draws from social-ecological systems frameworks to address urban planning and design problem domains. In recent work she has used this approach to investigate microclimate regulation through New Urbanist design, water, and biodiversity management through homeowners associations, and stormwater management through green infrastructure interventions. Dr. Turner's training is highly interdisciplinary.

Her work is funded by the National Science Foundation and the interdisciplinary National Socio-Environmental Synthesis Center. She recently chaired the Human Dimensions of Global Change specialty group of the American Association of Geographers. Dr. Turner deploys interdisciplinary pedagogy in the classroom and teaches courses in environmentalism, urban sustainability, and urban ecology. She received a Ph.D. in geography from the School of Geographical Sciences and Urban Planning at Arizona State University, where she was an IGERT Fellow in urban ecology.

KONGJIAN YU has been a professor of urban and regional planning and landscape architecture since 1997, and was the founding dean of the College of Architecture and Landscape at Peking University. He is the founder and design principal of Turenscape. His pioneering research on ecological security patterns and sponge city has been adopted by the Chinese government as guiding theory for a nationwide ecological urbanism campaign, and has had significant impact on national environmental policies in China. Dr. Yu has published over 20 books and 300 articles. His work has been featured in publications such as *Scientific American* (December 2018), *Landscape Architecture Magazine* (February 2012), and two recently published books: *Letters to Chinese Leaders: Kongjian Yu and the Future of the Chinese City* and *Designed Ecologies: The Landscape Architecture of Kongjian Yu*. His ecological approach to urbanism has been implemented in over 200 cities in China and abroad. His projects won numerous international awards including 12 ASLA Excellence and Honor Awards, 5 WAF World Best Landscape of the Year Awards, and a ULI global award of excellence. He is founder and chief editor of the internationally awarded magazine *Landscape Architecture Frontier*. He has lectured worldwide, including over 60 keynote speeches at international conferences, and taught for 5 years at the Harvard Graduate School of Design as a visiting professor. He was elected International Honorary Member of the American Academy of Arts and Sciences, fellow of the American Society of Landscape Architects, and received the Doctor Honoris Causa in Landscape and Environment from the Sapienza University of Rome and Honorary Doctor in Landscape Architecture from Norwegian University of Life Sciences. He received a Doctor of Design degree at the Harvard Graduate School of Design.

YAPING ZHANG is the vice president of CAS. Earlier in his career, he was a postdoctoral fellow at the Zoological Society of San Diego, Center for Reproduction of Endangered Species, until he went back to China and worked as the director of the Laboratory of Cellular and Molecular Evolution and a professor at the Kunming Institute of Zoology (KIZ). In 2002, Dr. Zhang was appointed professor and head of the Laboratory of Genetics at Yunnan University. He was nominated vice-president of CAS in 2012. As a research professor at KIZ, he has been focusing his research on molecular evolution and genome biodiversity. His investigations involve five correlated areas: molecular phylogenetics; molecular ecology and conservation genetics; human genetics and evolution; origin of domestic animals and artificial selection; and genome diversity and evolution. Dr. Zhang has published more than 300 publications in SCI journals and is the author and co-author of five books. He was elected a member of the American Society of Human Genetics in 1996, Society for Molecular Biology and Evolution in 1997, and American Genetic Association in 1998. He was also elected vice president of the Chinese Society of Genetics in 2004, vice president of the Chinese Society of Zoology in 2005, and president of the Yunnan Association for Science and Technology in 2008. Dr. Zhang sits on editorial boards for several international periodicals, including *Genome Biology and Evolution* and *Animal Genetics*. Dr. Zhang has won dozens of natural science prizes in China, including the Ho Leung Ho Lee Prize for Science and Technology by the Ho Leung Ho Lee Foundation in 2004. He was elected a CAS member in 2003 and fellow of the Third World Academy of Sciences in 2007. Dr. Zhang received his bachelor's degree from the Fudan University and his Ph.D. from the Kunming Zoology Institute of CAS.

WEIQI ZHOU is a professor of urban ecology at the Research Center for Eco-Environmental Sciences, Chinese Academy of Sciences. He is the deputy director of the State Key Laboratory of Urban and Regional Ecology and the director of the Beijing Urban Ecosystem Research Station, a long-term ecological research station focusing on the Beijing urban ecosystem. He is also the vice president of the Society for Urban Ecology (SURE), China Chapter. Before he joined the Research Center for Eco-Environmental Sciences, he was a postdoctoral fellow at the University of California, Davis. Dr. Zhou is broadly interested in urban and landscape ecology with respect to spatial heterogeneity of the landscape. He integrates field observations, remote sensing, and modeling to understand the structure of urban socio-ecological systems, and its link to ecological function. He works across many disciplines including landscape ecology, urban ecology, remote sensing, and GIS, and interacts with collaborators from different fields through his involvement with various collaborative projects. The interdisciplinarity of his work has allowed him to develop innovative approaches and tools to better understanding the structure of urban socio-ecological systems, and its link to ecological function. He serves as an editorial board member for the journals *Landscape Ecology, Landscape and Urban Planning, Journal of Urban Ecology, Remote Sensing, and Ecosystem Health* and *Sustainability*. Dr. Zhou received his Ph.D. from the University of Vermont.

YONG-GUAN ZHU (Planning Committee Member), professor of biogeochemistry and environmental biology, is the director-general of the Institute of Urban Environment, CAS. He has been working on the biogeochemistry of nutrients, metals, and emerging pollutants (such as antibiotics and antibiotic resistance genes). Dr. Zhu is a leader in taking multiscale and multidisciplinary approaches to soil and environmental problems. Before returning to China in 2002, he worked as a research fellow (supported by the Royal Society London), the Queen's University of Belfast, UK (1994–1995) and a postdoctoral fellow at the University of Adelaide (1998–2002), Australia. Dr. Zhu is currently the co-editor-in-chief of *Environment International* (Elsevier), and editorial member for several other international journals. He is a scientific committee member for the ICSU program on Human Health and Wellbeing in Changing Urban Environment, and served for 9 years as a member of the Standing Advisory Group for Nuclear Application, International Atomic Energy Agency (2004–2012). Dr. Zhu is the recipient of many international and Chinese merit awards, including the TWAS Science Award (2013) and National Natural Science Award (2009). Dr. Zhu has published over 300 papers in international journals, and these publications have attracted over 20,000 citations (Web of Science) with an H-index of 79. He was selected as a Web of Science Highly Cited Researcher (2016, 2017, and 2018) and is an elected fellow of the American Association for the Advancement of Science. He obtained his B.S.c. from Zhejiang Agricultural University in 1989, M.S.c. from CAS in 1992, and Ph.D. in environmental biology from Imperial College, London, in 1998.

Appendix C

Registered Workshop Participants

Karen Seto (NAS) (*Chair*)
Yale University

Benjamin Averbuch
National Council for Science and the Environment

Jeannie Bellina
The George Washington University

Lisa Benton-Short
The George Washington University

Luis Bettencourt
University of Chicago

Anthony Gad Bigio
The George Washington University

Linda Blevins
National Science Foundation

Uwe Brandes
Georgetown University

Bryan Brooks
Baylor University

William Bruner
Tandem Consulting

Jianming Cai
Institute of Geographical Sciences and Natural Resources Research
Chinese Academy of Sciences

Franklin Carrero-Martínez
National Academies of Sciences, Engineering, and Medicine

Robert Cervero
University of California, Berkeley

Wei-Qiang Chen
Institute of Urban Environment
Chinese Academy of Sciences

Chanda Chhay
Caset Associates, LTD

Andre Coelho
Federal Rural University of Rio de Janeiro

E. William Colglazier
American Association for the Advancement of Science

Frances Colón
Jasperi Consulting

Shenghui Cui
Institute of Urban Environment
Chinese Academy of Sciences

Chengbin Deng
State University of New York at Binghamton

Chengri Ding
University of Maryland

Richard Esposito
Bureau of Labor Statistics

Tracy Evans
ASE360 Consulting

Yingling Fan
University of Minnesota

Melissa Franks
National Academies of Sciences, Engineering, and Medicine

Warren Friedman
U.S. Department of Housing and Urban Development

Matei Georgescu
Arizona State University

Bruce Hamilton
National Science Foundation

Huili Han

Susan Hanson (NAS)
Clark University

Olivia Harp
University of the District of Columbia

Marccus Hendricks
University of Maryland

Henry Hettger
National Archives and Records Administration

Chunlin Huang
Northwest Institute of Eco-Environment and Resources
Chinese Academy of Sciences

Kangning Huang
Yale University

APPENDIX C

Sun Hui
Bureau of International Cooperation
Chinese Academy of Sciences

Yuelei Jin
Xinhua News Agency

Emi Kameyama
National Academies of Sciences,
  Engineering, and Medicine

William Kelly (NAE)
George Mason University

Jiangtao Li
National Natural Science
  Foundation of China

Tao Lin
Institute of Urban Environment
Chinese Academy of Sciences

Christopher Lindsay

Kejia Liu
Embassy of the People's Republic
  of China

Jose Lobo
Arizona State University

Richa Mahtta
Yale University

Dan Martinez
Tritico & Rainey

Margaret Matter
Oregon Department of Agriculture

Robert McDonald
The Nature Conservancy

Marcia McNutt (NAS)
President
U.S. National Academy of Sciences

Rajesh Mehta
National Science Foundation

Fanxin Meng
Yale University

Jerry Miller
Science for Decisions LLC

Cherry Murray (NAS/NAE)
University of Arizona/
  The InterAcademy Partnership

Deb Niemeier (NAE)
University of Maryland

Jiahua Pan
Chinese Academy of Social Sciences

Ashley Pierce
National Science Foundation

Nazanin Rezapour
Architecture Designer

Miguel Román
Universities Space Research
  Association

Jennifer Saunders
Consultant

Brandi Schottel
National Science Foundation

Yan Song
University of North Carolina

Teresa Stoepler
National Academies of Sciences, Engineering, and Medicine/ The InterAcademy Partnership

Eleanor Stokes
National Aeronautics and Space Administration

Zhongchang Sun
Institute of Remote Sensing and Digital Earth
Chinese Academy of Sciences

Bob Tansey
The Nature Conservancy

Amber Todoroff
Environmental and Energy Study Institute

Ting Tong
Bureau of International Cooperation
Chinese Academy of Sciences

V. Kelly Turner
University of California, Los Angeles

Enkhtaivan Urcheen
George Mason University

John Varljen
Varljen Consulting

Zhenyu Wang
Bureau of International Cooperation
Chinese Academy of Sciences

Zhongcheng Wang
Embassy of the People's Republic of China

Nathaniel Warren

Gang Wu
National Natural Science Foundation of China

Kongjian Yu
Peking University

Kamran Zendehdel
University of the District of Columbia

Yaping Zhang
Vice President
Chinese Academy of Sciences

Weiqi Zhou
Research Center for Eco-Environment Studies
Chinese Academy of Sciences

Yong-Guan Zhu
Institute of Urban Environment
Chinese Academy of Sciences